Behavior and
Design of
Aluminum
Structures

Behavior and Design of Aluminum Structures

Maurice L. Sharp

McGraw-Hill, Inc.

New York   St. Louis   San Francisco   Auckland   Bogotá
Caracas   Lisbon   London   Madrid   Mexico   Milan
Montreal   New Delhi   Paris   San Juan   São Paulo
Singapore   Sydney   Tokyo   Toronto

Library of Congress Cataloging-in-Publication Data

Sharp, Maurice L.
    Behavior and design of aluminum structures / Maurice L. Sharp.
        p.   cm.
    Includes index.
    ISBN 0-07-056478-7
    1. Aluminum construction. 2. Structural design. I. Title.
TA690.S44   1992
624.1'826—dc20   92-22257
                                                          CIP

1 2 3 4 5 6 7 8 9 0 DOC/DOC 9 8 7 6 5 4 3 2

ISBN   0-07-056478-7

The sponsoring editor for this book was Larry S. Hager, the editing supervisor was Stephen M. Smith, and the production supervisor was Donald F. Schmidt. It was set in Century Schoolbook by McGraw-Hill's Professional Book Group composition unit.

Printed and bound by R. R. Donnelley & Sons Company.

Contents

Preface

In over 100 years of commercial production of aluminum and its use in various finished products, a great deal of experience and research has been accumulated in the United States and throughout the world related to the behavior of aluminum structures. Some of this research is reflected in The Aluminum Association publications in the United States, but, in general, much useful information for design is scattered in various publications, technical journals, and company internal reports. The purpose of this book is to collect and interpret published and unpublished information that might be of interest to the structural designer of aluminum products. Data on problems and failures that have occurred in various finished products during service are provided to emphasize the approximate nature of the design process and the difficulty in anticipating all important considerations.

This book is intended to serve as a reference document for structural design engineers and students in advanced design courses with an interest in and a need for design information about aluminum. The emphasis is on the application of the science of aluminum to practical problems. The information is applicable to all aluminum structures.

The design of aluminum structures usually is not taught in the universities to the extent that other construction materials are covered. Thus, background information on such topics as production of the metal, product forms, alloy designations, alloy selection, typical applications, and design considerations is provided to give the designer a better understanding of the merits and limitations of aluminum. Also, some limited discussion is included on the similarities of and differences between aluminum and steel as they might affect design.

Laboratory data and field experience are used as much as possible to illustrate the accuracy and, in some cases, the uncertainty of the available analyses and design procedures. In many problems, testing and/or computer structural simulations are suggested as appropriate and sometimes necessary steps to verify the design of important structures.

Maurice L. Sharp

Acknowledgments

Much of the research that is incorporated in this book was conducted by Alcoa Laboratories scientists over the past 30 to 40 years. Some of the contributors are obvious because they are the authors of publications given in the references. Others are not so obvious but they also have contributed excellent work in unpublished reports or in supporting others in research. A significant amount of the information contained in this book is the result of people in the latter category. Contributions included in this book by other researchers in the world also are acknowledged.

My thanks are to my friends and family who encouraged me to write this book. My special thanks are to Dr. John W. Clark, retired scientist from Alcoa, for reading the drafts of chapters and making many constructive suggestions, and for his encouragement to me on this project.

Behavior and Design of Aluminum Structures

1

The Metal Aluminum

1.1 History

Aluminum, of all the metals now used for broad structural applications, was the last to come into widespread use. It is also the most abundant. Although it makes up one-twelfth of the earth's crust, aluminum was essentially a precious metal prior to the late nineteenth century: for example, about 50,000 lb (23,000 kg) were produced in the United States in 1891. In contrast, other common metals have been utilized for thousands of years: The Egyptians were making wrought iron in 1500 B.C. Blast furnaces existed in Europe in 1340. Copper has been in use for 5000 years.

The reason for the late development of large-scale uses for our most abundant metallic element is simple. Aluminum does not occur in the free state, but in combination with oxygen and other elements. These compounds of aluminum are very stable and require high temperature and energy to reduce them to metal. Thus, a more difficult process was needed to produce aluminum compared to that for many other metals, and its development took more time. The same property that accounts for aluminum's late entry into the marketplace gives the metal certain advantages in many present-day applications. The rapidity with which aluminum reacts with oxygen is an advantage in that a hard oxide layer forms on aluminum products that protects them from general atmospheric attack. Scratches heal quickly. This same oxide layer is of some disadvantage in welding, however. It must be broken up because it has a much higher melting point than aluminum.

Practical techniques for commercial production of aluminum date back to the nearly simultaneous discovery of an electrolytic process by Charles Martin Hall of the United States and Paul L. T. Héroult of France in 1886. Hall was able to prove that he had reduced his invention to practice 2 months before Héroult, and thus gained the patent rights in the United States.

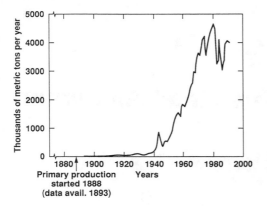

Figure 1.1 Primary production of aluminum in the United States.[1,2,3]

Growth of aluminum production was rapid after Hall's invention. Figure 1.1 shows that primary aluminum production in the United States increased rapidly, particularly between the 1940s and the 1970s. The fluctuations shown in Fig. 1.1, particularly in the 1980s, reflect effects of recycling and offshore production. Figure 1.2 shows data based on total supply of metal in the United States. The production of aluminum continues to grow.

Another measure of growth and size of the industry is revenues of the producing companies. In the initial 100 years (plus), revenues of the aluminum industry in North America have grown from nothing to over $30 billion (1990), rivaling the revenues generated by the more mature steel industry in the United States.

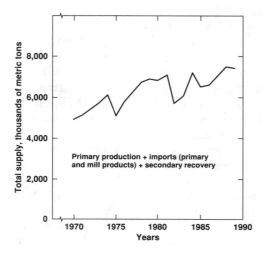

Figure 1.2 Total supply of primary aluminum in the United States.[1,2]

1.2   The Ore

Aluminum is extracted from aluminum oxide, or alumina. There are many deposits that contain aluminum oxides, the most economical being bauxite, a reddish ore that is most abundant in the equatorial regions of the world. At this time, there is no significant mining of bauxite in the United States. Bauxite deposits are usually covered by relatively thin layers of soil, so that open-pit mining is feasible.

The principal constituents of the bauxites currently mined are aluminum oxide, Al_2O_3; silicon dioxide, SiO_2; iron oxide, Fe_2O_3; and titanium dioxide, TiO_2. The specific composition depends on the particular deposit. Recent major producers of bauxite, their production, and estimated reserves are given in Table 1.1. Obvious from this table is that the recent major sources of high-grade bauxite are Australia, Brazil, Jamaica, and Guinea and that the total reserve base for all producing countries is hundreds of years. Lesser-grade ore can also be used if needed in the distant future. Although bauxite is the only raw material used in the production of alumina on a commercial scale, there are other sources that could satisfy primary aluminum and other alumina-based products needs.[3] Extraction from vast deposits of clays in the United States is technically feasible; such raw materials as anorthosite, alunite, coal wastes, and oil shales offer additional po-

TABLE 1.1   World Bauxite Production and Reserves
(In thousands of metric tons)

Country	Mine production			Reserve base[3]
	1983[4]	1986[5]	1990[3]	1990
Australia	23,000	32,400	39,600	4,600,000
Brazil	3,900	6,400	7,900	2,900,000
Greece	3,100	2,200	2,500	650,000
Guinea	9,000	14,700	17,500	5,900,000
Guyana	800	2,100	1,300	900,000
Hungary	2,800	—	2,600	300,000
India	2,000	2,300	4,300	1,200,000
Jamaica	7,400	7,000	10,900	2,000,000
Surinam	2,700	3,700	3,500	600,000
United States	700	500	—	40,000
U.S.S.R.	4,600	—	4,600	300,000
Yugoslavia	3,600	3,500	3,100	400,000
Other market economy countries	5,000	—		
			9,400	4,700,000
Other centrally planned economies	2,300	—		
Total	70,900	—	107,200	24,490,000

tential alumina sources. Thus, there is little likelihood of exhausting available sources, even if the growth of aluminum usage continues. Also, the deposits are well distributed in the world, so that a long-term disruption of supply for political reasons is unlikely to occur. An additional significant factor is that aluminum is easily recycled; the metal can be incorporated into new products after the original product is no longer useful. The net effect of such recycling is that fewer tons of ore will need to be mined and less metal will need to be smelted to sustain current applications and introduce new ones.

1.3   Production of the Metal

Pure aluminum oxide, Al_2O_3, is used for aluminum production. The aluminum oxide is obtained from bauxite by the Bayer process.[4] The bauxite entering the refinery is made uniform by blending from several sources and by crushing and grinding hard particles. The important part of the refining process then follows: the aluminum compounds are dissolved in NaOH solutions at elevated temperature; most other components of the bauxite do not dissolve, and can be removed. Refineries for the production of aluminum oxide are located in many countries as illustrated in Table 1.2.

TABLE 1.2   **Major Nonsocialist World Alumina Producers**[*,5]
(Production in thousands of metric tons)

Country	Annual production (mid-1986 to mid-1987)
Australia	9590
Brazil	1200
Canada	1200
France	1100
Greece	600
Guinea	700
India	1600
Ireland	800
Italy	720
Jamaica	1895
Japan	975
Spain	800
Surinam	1400
Turkey	200
United Kingdom	120
United States	3750
West Germany	1450
Yugoslavia	1300

*Countries designated as nonsocialist in 1988.

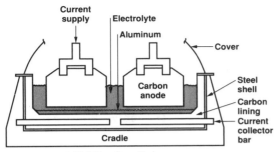

Figure 1.3 Hall-Héroult cell.

The metal is produced by electrolysis[6] in a Hall-Héroult cell of the type illustrated in Fig. 1.3. The cells are 30 to 40 ft (9 to 12 m) long, 10 to 13 ft (3 to 4 m) wide, and 3 to 4 ft (1 to 1.2 m) high, and are constructed of steel lined with carbon. The electrolyte is a solution of aluminum oxide in a molten salt solution composed primarily of a mixture of sodium fluoride and aluminum fluoride. The electrical current flows from the carbon anodes through the electrolyte to the carbon liner and the collection bars. The carbon anodes are consumed during the electrolysis. The aluminum collects at the bottom of the cell and is removed. The cells operate at a temperature of 1700 to 1800°F (930 to 980°C). Theoretically, 1 kAh of electric current will produce 0.74 lb (0.336 kg) of aluminum. The cells produce about 85 to 95 percent of this amount. The metal from the smelting pots is cast into primary ingots for subsequent remelting and use or is alloyed and cast into suitably shaped ingots for subsequent fabrication into various product forms. Almost all aluminum is alloyed to obtain desired characteristics, such as strength, corrosion resistance, and toughness. Table 1.3

TABLE 1.3 World Primary Aluminum Production[1,2]

(In thousands of metric tons)

Region	1980	1985	1988
Africa	437	513	541
North America	5,722	4,782	5,479
Latin America	821	1,157	1,521
East Asia	1,174	245	55
South Asia	386	902	912
Europe	3,747	3,596	3,762
Oceania	458	1,092	1,350
Other	2,638	3,080	3,684
Total	15,383	15,367	17,304

provides some recent statistics on the primary production of alumi-num in the world. North America continues to account for over 30 per-cent of the world's production.

1.4 Product Forms

Aluminum is available in all of the forms common with other metals: flat-rolled products, extrusions, forgings, and castings.[7,8] In all of the product forms, the quality of the part and its mechanical and corrosion properties depend not only on the chemistry of the alloy but on the process used to make the product. Thus, careful control is required in the manufacturing stage.

In flat-rolled products, thicknesses less than 0.006 in (0.15 mm) are referred to as *foil,* those equal to or greater than 0.25 in (6.35 mm) are *plate,* and all thicknesses in between are *sheet.* Flat-rolled products are produced in rolling mills. The cylindrical rolls primarily reduce thickness and increase length of the ingot. Initial rolling is done with the metal hot; final passes can be done hot or cold depending on the final product. Thermal treatments are usually needed after rolling to obtain the desired properties.

Capacities of rolling facilities vary, but for some alloys plate is pro-duced in thicknesses of 7 in (175 mm) or more. Rolling mills are in operation that are capable of producing plate in widths up to 17 ft (5.3 m), although smaller widths are more common, and less expensive. Length generally does not govern the size of sheet or plate available; the size is limited by the size of the ingot. The maximum weight of a single plate produced is about 7900 lb (3600 kg), and the maximum weight of sheet supplied in coils is about 9900 lb (4500 kg). The sheet and plate can later be formed or fabricated into more complex three-dimensional shapes.

In recent years, there has been a rapid growth of mini-mills for the manufacture of flat-rolled aluminum products.[9] Continuous slab cast-ers, for example, require lower initial capital than the traditional rolling-mill complex. They often utilize scrap feedstock and concen-trate on a specific market. The mini-mills cast the alloys to ½-in (12.7-mm) thicknesses or less, avoiding much of the hot rolling needed with conventional mills. The mini-mills provide a low-cost product that has usefulness in some markets (building products, for example). The growth of this product in the United States is from a negligible amount produced in 1977 to about 20 percent of the total flat-rolled production in 1990. (This product made up also about 20 percent of to-tal worldwide production of flat-rolled aluminum in 1990.)

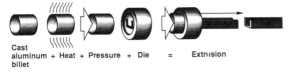

Cast
aluminum + Heat + Pressure + Die   =   Extrusion
billet

Figure 1.4   The extrusion process.

Extrusion is economical for parts with a complex cross section that is constant over its length. The process consists of pushing the metal from an extrusion ingot, or *billet,* through a die shaped to provide the finished configuration (see Fig. 1.4). Most alloys can be extruded although there are significant differences between the alloys in their ease of fabrication, and thus in finished costs. The largest extrusion presses in the industry are capable of applying a force of 18,000 tons (16,100 t) on the ingot although most of the presses are much smaller. Available gross ingot weights range up to 6800 lb (3090 kg). The large presses can produce flat sections up to 31 in (785 mm) wide. Cost of the final products varies with the size of the press and the alloy. The cross section of extrusions is constant along the length in most cases. Stepped extrusions have been made (the die needs to be changed during the extrusion), primarily for aircraft parts. The cross section of tubular extrusions can be changed along the length by spinning after extrusion; the tapered aluminum light pole is an example. Also, extrusions can be bent and twisted to obtain a desired longitudinal contour.

The forging process produces three-dimensional shapes by hammering or pressing the metal in a mold. Die forging is most common, with closed dies being employed. The forging ingot is placed in the cavity and the two parts of the die are pressed together at pressures of about 10 to 40 ton/in^2 (1.4 to 5.5 t/cm^2), forcing the metal to take the contour machined into the mold. The temperature of the metal being forged is about 730 to 890°F (390 to 480°C) depending on the alloy.

The largest forging press is 60,000 tons (54,000 t). Closed-die forgings 23 ft (7 m) long and weighing more than 3100 lb (1400 kg) have been produced. Much heavier open-die forgings (one die only) have been made. Most alloys can be forged, but those most commonly forged are medium- and high-strength alloys used for aerospace and wheel applications.

The casting process provides a very attractive way to achieve "net shape" (final shape of parts) three-dimensional configurations. The molten metal is simply introduced into a cavity configured essentially as the final part. There are many casting processes; the more common are die casting, permanent-mold casting, and sand casting. Differences lie in the type of mold used as well as the method by which the material is introduced into the mold. In die casting the metal is injected into reusable dies (steel) at high velocity. Rapid solidification occurs, so the process is capable of high production rates.

In permanent-mold casting the metal is poured into the cavity in the reusable molds, sometimes with a small vacuum to assist the flow. Generally, the sections are thicker than those in die casting, and solidification is slower.

In the sand-casting process the mold is destroyed each time, and solidification and cycle time are slower than for the other processes. However, this process is quite economical when the quantity of parts needed is low and the casting is large and complex in geometry.

There are numerous alloys used for castings, somewhat different in composition from those used for sheet, extrusions, and die forgings (wrought alloys), to accommodate the casting process. Size of castings is limited by the equipment. Generally, die-casting parts are the smallest, permanent molds are intermediate, and sand castings are the largest [220 lb (100 kg) or more]. Casting not only produces three-dimensional shapes that can be the final shape of the product, but also shapes that are much more intricate in geometry compared to the other product forms.

1.5 Alloy and Temper Designations

The Aluminum Association assigns alloy and temper designations in the United States. Wrought alloys (sheet and plate, extrusions, and forgings) have a four-digit number (see Table 1.4); cast alloys have a three-digit number to the left of the decimal point and one digit to the right of the decimal point (see Table 1.5). The first digit defines the major alloying ingredient for both wrought and cast alloys. The major alloying ingredient is usually 5 percent or less (by weight) in wrought alloys, and the same or higher in cast alloys. Most of the alloys contain two to four other elements, but in a much smaller amount than

TABLE 1.4   Wrought Alloy Designation System[10]

Alloy	Major alloying ingredient
1XXX	99% minimum aluminum
2XXX	Copper
3XXX	Manganese
4XXX	Silicon
5XXX	Magnesium
6XXX	Magnesium and silicon
7XXX	Zinc
8XXX	Other elements
9XXX	Unused series

TABLE 1.5   Cast Aluminum Alloy Designation System[10]

Alloy	Major alloying ingredient
1XX.X	99% minimum aluminum
3XX.X	Silicon, with added copper and/or magnesium
4XX.X	Silicon
5XX.X	Magnesium
7XX.X	Zinc
8XX.X	Tin
9XX.X	Other element
6XX.X	Unused series

the major alloying ingredient. The alloying ingredients greatly increase strength in most cases and affect most of the other properties, such as ease of fabrication, corrosion resistance, and toughness.

In the wrought alloy designation, the last two digits in the 1XXX series give the minimum aluminum percentage above 99.00 percent. In the 2XXX through 9XXX groups, the last two digits specify the individual alloys registered for the series. The second digit designates a modification of the original alloy.

The alloy designation system is similar for castings. In this case, the second and third digits give the minimum aluminum percentage above 99.00 percent for the 1XX.X alloys. In the 2XX.X through 9XX.X groups, the second two numbers are the individual alloys registered in the series. The number to the right of the decimal point gives product form; 0 for castings and 1 and 2 (narrower composition limits than 1) for ingot.

Table 1.6 provides the basic temper designation system. The temper tells how the product was fabricated, and applies to both wrought and cast products. F and O tempers apply to all alloys and product forms. Tempers designated -TXXXX apply to alloys and product forms that receive and respond to a thermal treatment after fabrication. These alloys are said to be *heat-treatable*. Wrought alloys in the 2XXX, 6XXX, and 7XXX series and the cast alloys are generally in this group. *Non-heat-treatable* alloys gain their strength and other characteristics by strain hardening, and a temper of -H is specified. Included in this group are the 1XXX, 3XXX, and 5XXX series. More details on tempers are contained in various Aluminum Association reference documents.[10]

Other countries have alternative designations for the alloys and tempers. Table 1.7 correlates some of the designations and alloys.

Table 1.8 presents physical properties of a few alloys. These characteristics of aluminum are typical properties and, as such, are not guaranteed.

TABLE 1.6 Temper Designations[10]

Temper	Fabrication treatments
F	*As fabricated.* Applies to the products of shaping processes in which neither special control over thermal conditions nor strain-hardening is employed. For wrought products, there are no mechanical property limits.
O	*Annealed.* Applies to wrought products which are annealed to obtain the lowest strength temper, and to cast products which are annealed to improve ductility and dimensional stability. The O may be followed by a digit other than zero.
H	*Strain-hardened (wrought products only).* Applies to products which have their strength increased by strain-hardening, with or without supplementary thermal treatments to produce some reduction in strength. The H is always followed by two or more digits.
H1	*Strain-hardened only.* Applies to products which are strain-hardened to obtain the desired strength without supplementary thermal treatment. The number following this designation indicates the degree of strain-hardening.
H2	*Strain-hardened and partially annealed.* Applies to products which are strain-hardened more than the desired final amount and then reduced in strength to the desired level by partial annealing. For alloys that age-soften at room temperature, the H2 tempers have the same minimum ultimate tensile strength as the corresponding H3 tempers. For other alloys, the H2 tempers have the same minimum ultimate tensile strength as the corresponding H1 tempers and slightly higher elongation. The number following this designation indicates the degree of strain-hardening remaining after the product has been partially annealed.
H3	*Strain-hardened and stabilized.* Applies to products which are strain-hardened and whose mechanical properties are stabilized either by a low temperature thermal treatment or as a result of heat introduced during fabrication. Stabilization usually improves ductility. This designation is applicable only to those alloys which, unless stabilized, gradually age-soften at room temperature. The number following this designation indicates the degree of strain-hardening before the stabilization treatment.
W	*Solution heat-treated.* An unstable temper applicable only to alloys which spontaneously age at room temperature after solution heat treatment. This designation is specific only when the period of natural aging is indicated (for example, W ½ h).
T	*Thermally treated to produce stable tempers other than F, O, or H.* Applies to products which are thermally treated, with or without supplementary strain-hardening, to produce stable tempers. The T is always followed by one or more digits.
T1	*Cooled from an elevated-temperature shaping process and naturally aged to a substantially stable condition.* Applies to products which are not cold-worked after cooling from an elevated-temperature shaping process, or in which the effect of cold work in flattening or straightening may not be recognized in mechanical property limits.

TABLE 1.6   Temper Designations[10] (*Continued*)

Temper	Fabrication treatments
T2	*Cooled from an elevated-temperature shaping process, cold-worked, and naturally aged to a substantially stable condition.* Applies to products which are cold-worked to improve strength after cooling from an elevated-temperature shaping process, or in which the effect of cold work in flattening or straightening is recognized in mechanical property limits.
T3	*Solution heat-treated, cold worked, and naturally aged to a substantially stable condition.* Applies to products which are cold-worked to improve strength after solution heat treatment, or in which the effect of cold work in flattening or straightening is recognized in mechanical property limits.
T4	*Solution heat-treated and naturally aged to a substantially stable condition.* Applies to products which are not cold-worked after solution heat treatment, or in which the effect of cold work in flattening or straightening may not be recognized in mechanical property limits.
T5	*Cooled from an elevated-temperature shaping process and then artificially aged.* Applies to products that are not cold-worked after cooling from an elevated-temperature shaping process, or in which the effect of cold work in flattening or straightening may not be recognized in mechanical property limits.
T6	*Solution heat-treated and then artificially aged.* Applies to products which are not cold-worked after solution heat treatment, or in which the effect of cold work in flattening or straightening may not be recognized in mechanical property limits.
T7	*Solution heat-treated and overaged/stabilized.* Applies to wrought products that are artificially aged after solution heat treatment to carry them beyond a point of maximum strength to provide control of some significant characteristic. Applies to cast products that are artificially aged after solution heat treatment to provide dimensional and strength stability.
T8	*Solution heat-treated, cold-worked, and then artificially aged.* Applies to products which are cold-worked to improve strength, or in which the effect of cold work in flattening or straightening is recognized in mechanical property limits.
T9	*Solution heat-treated, artificially aged, and then cold-worked.* Applies to products which are cold-worked to improve strength.
T10	*Cooled from an elevated-temperature shaping process, cold-worked, and then artificially aged.* Applies to products that are cold worked to improve strength, or in which the effect of cold work in flattening or straightening is recognized in mechanical property limits.

TABLE 1.7 Alloy Designations[10]

Alloy designation	Designating country	Equivalent or similar AA alloy
Al99		1200
Al99.5		1050
E-Al		1350
AlCuMg1		2017
AlCuMg2	Austria	2024
AlCuMg0.5	(Önorm)[a]	2117
AlMg5		5056
AlMgSi0.5		6063
E-AlMgSi		6101
AlZnMgCu1.5		7075
990C		1100
CB60		2011
CG30		2117
CG42		2024
CG42 Alclad		Alclad 2024
CM41		2017
CN42		2018
CS41N		2014
CS4N Alclad		Alclad 2014
CS41P		2025
GM31N		5454
GM41	Canada	5083
GM50P	(CSA)[b]	5356
GM50R		5056
GR20		5052
GS10		6063
GS11N		6061
GS11P		6053
MC10		3003
S5		4043
SG11P		6151
SG121		4032
ZG62		7075
ZG62 Alclad		Alclad 7075
A5/L		1350
A45		1100
A-G1		5050
A-G0.6		5005
A-G4MC		5086
A-GS		6063
A-GS/L		6101
A-M1	France	3003
A-M1G	(NF)[c]	3004
A-U4G		2017
A-U2G		2117
A-U2GN		2618
A-U4G1		2024
A-U4N		2218
A-U4SG		2014

TABLE 1.7 Alloy Designations[10] (Continued)

Alloy designation	Designating country	Equivalent or similar AA alloy
A-S12UN	France	4032
A-Z5GU	(NF)[c]	7075
E-A1995[d] (3.0257[e])		1350
AlCuBiPb[d] (3.1655[e])		2011
AlCuMg0.5[d] (3.1305[e])		2117
AlCuMg1[d] (3.1325[e])		2017
AlCuMg2[d] (3.1355[e])		2024
AlCuSiMn[d] (3.1255[e])	Germany	2014
AlMg4.5Mn[d] (3.3547[e])		5083
AlMgSi0.5[d] (3.3206[e])		6063
AlSi5[d] (3.2245[e])		4043
E-AlMgSi0.5[d] (3.3207[e])		6101
AlZnMgCu1.5[d] (3.4365[e])		7075
1E		1350
91E		6101
H14		2017
H19		6063
H20		6061
L.80, L.81		5052
L.86		2117
L.87		2017
L.95, L.96	Great Britain	7075
L.97, L.98	(BS)[f]	2014
2L.55, 2L.56		5052
2L.58		5056
3L.44		5050
5L.37		2017
6L.25		2218
N8		5083
N21		4043
150A		2017
324A		4032
372B	Great Britain	6063
717, 724, 731A, 745, 5014, 5084	(DTD)[g]	2618
5090		2024
5100		Alclad 2024
P-AlCu4MgMn		2017
P-AlCu4.5MgMn		2024
P-AlCu4.5MgMnplacc.		Alclad 2024
P-AlCu2.5MgSi		2117
P-AlCu4.4SiMnMg	Italy	2014
P-AlCu4.4SiMnMgplacc.	(UNI)[h]	Alclad 2014
P-AlMg0.9		5657
P-AlMg1.5		5050
P-AlMg2.5		5052
P-ALSi0.4Mg		6063
P-AlSi0.5Mg		6101

TABLE 1.7   Alloy Designations[10] (Continued)

Alloy designation	Designating country	Equivalent or similar AA alloy
Al99.5E		1350
L-313		2014
L-314	Spain	2024
L-315	(UNE)[i]	2218
L-371		7075
Al-Mg-Si		6101
Al1.5Mg		5050
Al-Cu-Ni	Switzerland	2218
Al3.5Cu0.5Mg	(VSM)[j]	2017
Al4Cu1.2Mg		2024
Al-Zn-Mg-Cu		7075
Al-Zn-Mg-Cu-pl		Alclad 7075
Al99.0Cu		1100
AlCu2Mg		2117
AlCu4Mg1		2024
AlCu4SiMg		2014
AlCu4MgSi		2017
AlMg1		5005
AlMg1.5		5050
AlMg2.5		5052
AlMg3.5	ISO[k]	5154
AlMg4		5086
AlMg5		5056
AlMn1Cu		3003
AlMg3Mn		5454
AlMg4.5Mn		5083
AlMgSi		6063
AlMg1SiCu		6061
AlZn6MgCu		7075

[a]Austrian Standard M3430.
[b]Canadian Standards Association.
[c]Normes Françaises.
[d]Deutsche Industrie-Norm.
[e]Werkstoff-Nr.
[f]British Standard.
[g]Directorate of Technical Development.
[h]Unificazione Nazionale Italiana.
[i]Una Norma Español.
[j]Verein Schweizerischer Maschinenindustrieller.
[k]International Organization for Standardization.

TABLE 1.8 Typical Physical Properties[10,11]

Alloy and temper	Specific gravity	Density		Average coefficient of thermal expansion		Melting range		Thermal conductivity		Electrical conductivity, % of international annealed copper standard		Electrical resistivity	
		lb/in³	kg/m³ ×10³	68–212°F ×10⁶	20–100°C, ×10⁶	°F	°C	at 77°F, (Btu·in)/(ft²·h·°F)	at 25°C, W/(m·K)	Equal volume	Equal weight	at 68°F, (Ω·cmil)/ft	at 20°C, (Ω·mm²/m)
1100-0	2.71	0.098	2.71	13.1	23.6	1190–1215	640–655	1540	222	59	194	18	0.029
2014-T6	2.80	0.101	2.80	12.8	23.0	945–1180	505–635	1070	155	40	127	26	0.043
2024-T3	2.78	0.101	2.78	12.9	23.2	935–1180	500–635	840	121	30	96	35	0.057
2219-T81	2.84	0.103	2.84	12.4	22.3	1010–1190	545–645	840	121	30	94	35	0.057
3004-H32	2.72	0.098	2.72	13.3	23.9	1165–1210	630–655	1130	163	42	137	25	0.041
4043-0	2.68	0.097	2.68	12.3	22.0	1065–1170	575–630	1130	163	42	140	25	0.041
5052-H32	2.68	0.097	2.68	13.2	23.8	1125–1200	605–650	960	138	35	116	30	0.049
5083-0	2.66	0.096	2.66	13.2	23.8	1095–1180	580–640	810	117	29	98	36	0.059
5086-H116	2.66	0.096	2.66	13.2	23.8	1085–1185	585–640	870	126	31	104	33	0.056
5456-H116	2.66	0.096	2.66	13.3	23.9	1055–1180	570–640	810	117	29	98	36	0.059
6061-T6	2.70	0.098	2.70	13.1	23.6	1080–1205	580–650	1160	167	43	142	24	0.040
6063-T5	2.69	0.097	2.69	13.0	23.4	1140–1210	615–655	1450	209	55	181	19	0.031
7075	2.80	0.101	2.80	13.1	23.6	890–1175	475–635	900	130	33	105	31	0.052
356.0-T6	2.68	0.097	2.68	11.9	21.4	1035–1135	555–615	1040	150	39	129	—	—

For normal structural design of most products, average values of most physical properties may be used because they do not vary greatly with product form, alloy, or temper.

References

1. *Aluminum Statistical Review,* The Aluminum Association, Washington, D.C., 1989.
2. "Historical Supplement," *Aluminum Statistical Review,* The Aluminum Association, Washington, D.C., 1982.
3. *Mineral Commodity Summaries,* U.S. Bureau of Mines, 1991.
4. *Ullmann's Encyclopedia of Industrial Chemistry,* Vol. A1, "Aluminum Oxide," VCH Verlagsgesellschaft mbH, D-6940, Weinheim, West Germany, 1985, p. 557.
5. *Annual Review of the World Aluminium Industry, 1988,* Shearson Lehman Brothers Limited, London, February 1988.
6. *Ullmann's Encyclopedia of Industrial Chemistry,* Vol. A1, "Aluminum Alloys," VCH Verlagsgesellschaft mbH, D-6940, Weinheim, West Germany, 1985, p. 459.
7. *Ullmann's Encyclopedia of Industrial Chemistry,* Vol. A1, "Aluminum Alloys," VCH Verlagsgesellschaft mbH, D-6940, Weinheim, West Germany, 1985, p. 481.
8. Van Horn, Kent R. (ed.), *Aluminum,* Vol. 3, "Fabrication and Finishing," American Society for Metals, Metals Park, Ohio, 1967.
9. *Proceedings of Ingot and Continuous Casting Process Technology Seminar for Flat Rolled Products,* Vol. II, sponsored by Sheet and Plate Product Division, The Aluminum Association, New Orleans, Louisiana, May 10–12, 1989.
10. *Aluminum Standards and Data,* The Aluminum Association, Washington, D.C., 1990.
11. *Standards for Aluminum Sand and Permanent Mold Castings,* The Aluminum Association, Washington, D.C., March 1980.

2

Applications and Alloy Selection

2.1 Market Development

Markets for aluminum grew quickly after the discovery of the Hall process (see Table 2.1).[1] The earliest uses were for cooking utensils and electrical cable. The wars during the 1910s and 1940s consumed much of the metal produced for various products needed to support those efforts. The most recent significant market to develop is packaging, with beverage cans being the largest item.

Figure 2.1 shows the major markets in recent years.[2] Container and packaging markets include beverage cans, semirigid food containers, closures, and flexible packaging. The electrical, machinery and equipment, and consumer durable goods markets encompass many areas, such as wire and cable products, industrial machinery, pipe, ladders, appliances, and air conditioning products. Transportation applications include passenger cars, trucks, buses, ships, recreational vehicles, aircraft, and railcars. Uses in building and construction include siding and roofing products, curtain walls, gutters and downspouts, highway signs, sign structures, windows, and lighting poles. There has been a

TABLE 2.1 Development of Major Markets

Market	Approximate starting dates
Cooking utensils	1890s
Electrical cable	1900s
Military	1910s
Transportation	1920s, 1930s
Aircraft, military	1940s
Building products	1950s
Packaging	1960s

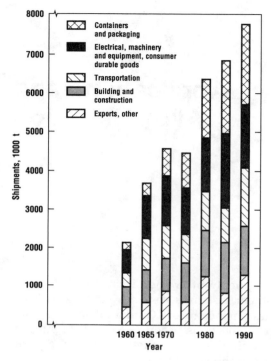

Figure 2.1 Shipments to major markets.[2]

continuing growth in the total market for aluminum products and increases in most of the market categories. The largest growth in recent years has been in the packaging market. In each of the market categories the use of aluminum has, in most cases, been at the expense of incumbent materials such as wood and steel.

2.2 Attributes of the Metal

The reasons for the use of a material in any application are many, but in the final analysis the product must be affordable: the cost of the product must be acceptable to the customer. The cost of various aluminum products, as well as the cost of aluminum relative to steel, is dealt with in some detail in subsequent sections. Generally, aluminum is attractive in many applications because of a favorable life-cycle cost—the sum of the initial cost of the finished product, the cost of operating or maintaining the product over its life, and the cost of disposing of or recycling it after its useful life. In addition, aluminum has sustained and increased its use in many markets partly because the price for aluminum relative to that for steel, overall, has decreased gradually over the aluminum industry's 100-year life. Some of the mechanical and physical characteristics of aluminum that have influenced market development follow.

- *Lightweight/high strength-to-weight ratio.* Aluminum has about one-third the density of steel and has alloys with strengths comparable to many of the constructional steels. This characteristic usually allows an aluminum design to weigh one-third to one-half that of a corresponding steel part, depending on design requirements. In aircraft, particularly commercial craft, aluminum is used extensively because the low weight cuts fuel requirements, the net effect being the most cost-effective design. Designs for other transportation vehicles—trucks, railroad coal cars, and passenger cars, for example—utilize aluminum for the same reason. Also, strength and toughness increase at low temperatures, making aluminum desirable for containers used in cryogenic applications.

- *Corrosion resistance.* The surface of bare aluminum oxidizes quickly in air. This oxide layer is hard and highly resistant to corrosion from industrial and seacoast atmospheres. Damaged spots self-heal. Also, paint and other finishes on aluminum stay attractive: Painted finished products for houses maintain their appearances over 20 years, and the finished panels on the Alcoa Building in Pittsburgh are attractive after 40 years. Light poles, painted signs, and many residential products make use of this characteristic. These applications generally have a long-life requirement for acceptable appearance.

- *Ease of fabrication.* Some assembly costs can be eliminated by use of the intricate final shapes possible from castings, extrusions, and forgings. The designer can place the material in locations to achieve maximum efficiency of the shape of the part with these product forms. In addition, fewer parts are needed for the aluminum product (compared to other materials) because the designer can include features in the shape that permit it to satisfy several functions. Window frame sections and curtain-wall support sections are examples of extrusions that have replaced assemblies of other competing materials.

   Sheet supplied for aircraft, automobiles, and packaging is capable of being formed into the shapes required for these applications. Some alloys can be superplastically formed to shape, which allows much more severe forming and a reduction in number of parts required.

   Normal metalworking and joining methods are established—cutting, machining, forming, welding, mechanical fastening, and adhesive bonding, for example. The methods generally are the same as for other metals and are no more difficult to accomplish than with other metals.

- *Desirable physical properties (nonmagnetic, nontoxic, high thermal and electrical conductivities, high reflectivity).* Some of the earliest applications for aluminum, i.e., cookware and electrical conductor cables, were developed because of the favorable thermal and electrical conductivity and cost of aluminum compared to the incumbent materials.

- *Ease of designing alloys for applications.* Different market areas and products have different requirements for strength, corrosion resistance, and toughness, and aluminum producers have responded with new alloys to fit application requirements. In effect, it is normal to design the alloy to meet the requirements of products in major markets, and the ease with which new alloys can be designed is a major advantage of aluminum as a material.

Obviously, many existing products capitalize on more than one of the attributes mentioned above. One question that should be asked when considering aluminum for a new application is whether one or more of these attributes are important for the product. If so, aluminum should be a viable candidate for a cost-effective solution.

2.3   Examples of Products and Alloy Selection

One of the good sources of information for alloy selection for new products is the experience with past and present aluminum products. The following examples are intended to provide broad background on alloys and tempers successfully employed in various applications. The alloys and tempers discussed here are not uniquely used, but rather are representative of those employed.

Packaging. One of the common products is household or industrial foil. Alloy 1145-0 is one of the alloys used. This product, in the annealed temper, has low strength so that it is easily folded to shape, retains the shape, provides a moisture and gas barrier, and is resistant to corrosion by the products stored. The same alloy, when formed into shapes in the -H19 temper, is appropriate for the semirigid foil containers used for temporary storage of various food products. Foils are often laminated to plastic films and paper for other containers.

The beverage can shown in Fig. 2.2 has two pieces, the body and a coated end. The bottom portion, which is drawn and ironed to shape, is alloy 3004. Considerations are good formability and high strength. The can end is seamed on the body and is made from alloy 5182. The tab is made of alloy 5042. Strength is important to withstand internal pressure, filling, and stacking loads. Aluminum is uniquely suited for the can end because it has the strength required to withstand these

Figure 2.2   Aluminum beverage can. (*Courtesy of Alcoa.*)

pressures but also can be effectively scored so that it is consistently easy to open.

Some food cans are shown in Fig. 2.3. Alloys used for various types of containers are 5352, 5042, and 5182. The can is ribbed to withstand the partial vacuum loads during processing. Otherwise, the loads are of the same type as those for beverage cans.

Transportation. The aircraft structure is primarily exterior skin sheets riveted to stiffeners (ribs, longerons), spars, and bulkheads (see Fig. 2.4). Light weight is essential; therefore, the highest-strength alloys available are employed.[3] Generally, the toughness and corrosion resistance of these alloys is not as good as for some of the alloys that have moderate strengths. Also, the cost of the high-strength alloys generally is higher than that of the lower-strength alloys used in other applications. The alloys for the skin sheets are Alclad 2024-T3, Alclad 2014-T6, and Alclad 7075-T6. *Alclad* means that there has been a thin, high-purity alloy rolled on and metallurgically bonded to the base alloy to improve corrosion resistance. The stiffeners may be

Figure 2.3   Aluminum food cans. (*Courtesy of Alcoa.*)

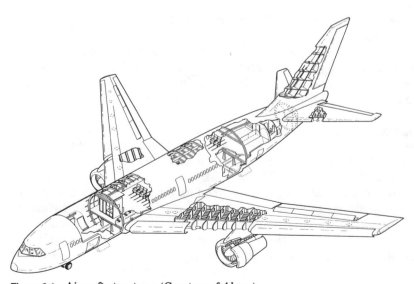

Figure 2.4   Aircraft structure. (*Courtesy of Alcoa.*)

formed from sheet, extruded, or forged. Alloys 7075, 7050, 7475, and 7178 are used. High-strength tempers are employed, -T6 or -T73, the latter case being an overaged temper that provides improved resistance to stress-corrosion cracking. Aluminum-lithium alloys and ARALL®, the latter a laminate of aircraft sheet and Kevlar mats, are new aluminum products with some commercial use.

The Alcoa *Seaprobe,* a 250-ft-long (76-m) vessel built and operated in the 1970s is shown in Fig. 2.5. The all-aluminum ship was designed to do oceanographic research, and there was an opening through the hull amidships directly under the 100-ft-tall (30-m) derrick so that various instruments could be lowered to the ocean depths.

The hull of the *Seaprobe* was constructed with longitudinal stiffeners and transverse frames and bulkheads. Alloy 5456-H116 was selected for the hull plating, to give high as-welded strength and good structural toughness. "Toughness" in this case meant the ability of the structure to accommodate plastic deformation without fracturing the material during minor collisions. Extrusions made of alloy 5456-H111 were used for the stiffeners—again chosen for toughness of the welded structure, even though their cost was much higher than that for 6061-T651 extrusions. The weld filler wire was alloy 5556. A strong, lightweight derrick was needed at reasonable cost, and tubu-

Figure 2.5   Alcoa *Seaprobe.*

lar members of 6061-T651 welded into a spaceframe using filler wire of alloy 5356 satisfied the requirements. Toughness and energy absorption were not primary requirements for the derrick. Spheres were employed at some nodes to facilitate fabrication.

Railroad cars (Fig. 2.6) have been another application of 5XXX alloy sheet and plate for good welded strength and toughness, particularly 5083-H113, and 5083-H321. Extrusions are generally 5083-H111, 5083-H112, or 6061-T6. Some cars have been built using riveted construction. Both methods of joining have produced acceptable cars. A common application for aluminum cars is to haul coal from mines to electrical power plants. The light weight of the cars allows higher payloads, making the application cost effective.

The use of aluminum in the passenger automobile has increased over the years. Until recently, most uses have been in hang-on components such as hoods and decklids to save weight (see Fig. 2.7). Alloys 6009-T4 and 2008-T4 are candidates for these parts because they provide the excellent formability needed to make the parts in the -T4 temper. In addition, with aging (subjecting the part to elevated temperature for a short time), the strength of the alloy is increased, thereby enhancing dent resistance and resistance to permanent set.

Figure 2.6   Aluminum railroad car. (*Courtesy of Alcoa.*)

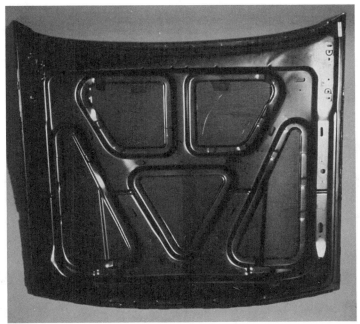

Figure 2.7   Inner panel of aluminum hood.

Sometimes the paint-bake cycle used for the automobile is sufficient to achieve this additional strength.

Considerable developmental work is now under way to utilize aluminum in the primary body of the automobile, and there has been limited production of all-aluminum cars. Figure 2.8 is an all-sheet concept that can be joined by spot welding, weld bonding, or riveting, and Fig. 2.9 is a concept that also employs 6XXX extrusions for the lineals, joined at nodes (ductile castings) by welding. The alloys and tempers are being developed for the castings and extrusions to closely fit the requirements of the application. The worldwide interest in increased use of aluminum in automobiles results from the desire to reduce fuel consumption and exhaust emissions. The weight of the aluminum body is about one-half the weight of a corresponding steel vehicle (when each is designed for the same requirements of safety, durability, and stiffness), which allows more fuel-efficient cars without sacrificing performance. In addition, aluminum is readily recycled (70 percent is recovered from the vehicle now) to make new parts from cars. The excellent corrosion resistance of aluminum is also an advantage. The current work is to develop concepts that are cost effective for aluminum producers, automobile manufacturers, and the customer.

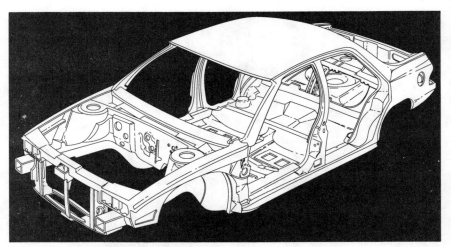

Figure 2.8 Sheet body-in-white.

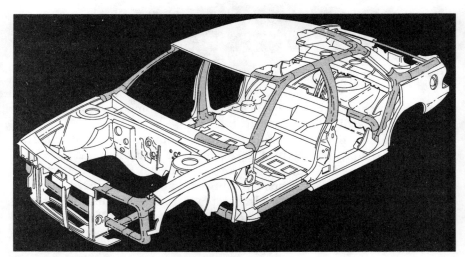

Figure 2.9 Mixed product body-in-white.

The application of extrusions and castings in addition to sheet is intended to help reduce manufacturing costs.

Building and construction. Figure 2.10 shows a recent photograph of the curtain wall installed on the Alcoa Building in 1952, one of the first such applications of aluminum on large buildings. The panels were 3003-H18 stampings and were treated with a special electrochemical bath to pro-

Figure 2.10 Curtain wall on Alcoa Building. (*Courtesy of Alcoa.*)

vide the permanent gray color and greater resistance to corrosion. Excellent service for 40 years has been obtained to date.

Uses of sheet aluminum in residential buildings include gutters and downspouts, roofing, siding, and fascia. Alloys 3003, 3105, 5050, and 5052 are representative. Extrusions for curtain-wall framing and window frames are 6063-T5, with some use of 6061-T6.

An overhead sign structure is shown in Fig. 2.11 after about 20 years exposure to a semi-industrial environment. The extruded tubes are generally 6061-T651. Splice flanges to join segments together and the base fixtures are castings of 356-T6. The parts are arc-welded together, using 4043 or 5356 fillers. These structures are utilized because they have an excellent appearance over many years without need for maintenance.

The light pole shown in Fig. 2.12 is another common use of 6063 extrusions. The extrusions are mechanically tapered to achieve the shape. The shaft is welded to a 356-T6 casting with 4043 filler. The

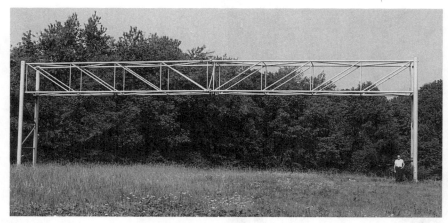

Figure 2.11 Aluminum overhead sign structure.

Figure 2.12 Aluminum light pole.

assembly is often aged after welding in the -T4 temper to improve as-welded strength. Consistent appearance without need for maintenance is the advantage of this product.

2.4 Alloy and Temper Selection

Table 2.2 provides some guidance on selection of alloy and temper for various product forms and requirements. Only a few of the common alloys and tempers are shown; aluminum producers and The Aluminum Association have more detailed information. The factors considered for an application are as follows:

- Product form
- Fabrication method
- Environment (corrosion, temperature, loading, etc.)
- Maintainability
- Cost

Three finished product areas are shown: aerospace, packaging, and "general purpose" (everything else). The emphasis of the aerospace application is high strength-to-weight ratio. Thus, the alloys have high strength, but moderate corrosion resistance and toughness. Careful design and testing of each structure is done to optimize design. Frequent inspections of operating aircraft are required to verify the adequacy of the structure. Repairs are made as needed. The aircraft companies have structural design technology especially developed for aircraft applications of aluminum.

The packaging industry makes use of only a few alloys and product forms. Important are alloys that have high recyclability and good corrosion resistance, are easily formed to shape, and have moderate strength. Household foil is annealed so that it can be easily formed to a shape required. Aluminum foil provides an excellent barrier to moisture or gas transmission.

The "general-purpose" alloys are employed in a wide range of other applications, and this category presents the greatest variety for alloy selection. This discussion covers the five considerations listed above.

Product form. All of the wrought products—foil, sheet, plate, extrusions, and forgings—are manufactured by the major aluminum producers, who obtain high quality and consistency in each product by careful control of chemistry and process. The properties of these products generally are well established and available from The Aluminum Association, other code and standards organizations, and the produc-

TABLE 2.2 Alloy and Temper Selection Chart

Product form	Aerospace (high strength, moderate toughness, moderate corrosion resistance)	Packaging, rigid and flexible	General purpose (low to moderate strength, highest corrosion resistance)	General purpose (readily welded)	General purpose (highest toughness in welded structure)
Sheet and plate	Alclad 2024-T3 6013-T6 Alclad 2014-T6 Alclad 7075-T6 7075-7651 2219-T851	3004-H19 5052-H19 5182-H19 3003	5052-H32 5454-H32 5083-H32 5086-H116 5086-H116 3003-H16 6061-T6 3004-H32	5052-H32 5454-H32 5083-H32 5086-H116 5456-H116 6061-T6	5083-0
Extrusion	7075-T73 7050-T73 6013-T6 7475-T73		6061-T651 6063-T5	6061-T651 6063-T5	5086-H111 5456-H111 6061-T4
Forging	7075-T73		6061-T6	6061-T6	
Casting	A356-T6		356-T6 A444-T4	356-T6 A444-T4	A444-T4
Foil		1145-0 1145-H19			

Product form	General purpose (sustained high temperature, <400°F)	General purpose (high formability)	General purpose (lowest cost per pound)	General purpose (good formability, moderate strength)	Ballistic performance alloys
Sheet and plate	5052-H32 5454-H32 6061-T6	5182-0 6009-T4 2008-T4 2036-T4	5052-H32	3003-H16 3004-H32	7049-T64 5083-H131
Extrusion	6061-T651 6063-T5	6XXX-T4	6063-T5		
Forging	6061-T6		6061-T6		
Casting	356-T6 A444-T4		356-T6		
Foil					

ers. Castings, on the other hand, are made by many large and small producers and can have a variety of characteristics and quality, ranging from carefully made and inspected aerospace parts to less critical parts that must meet a shape requirement but have few or no performance requirements. Because process variables influence properties, such as strength, corrosion resistance, and ductility, that determine final quality, the design of the casting must accommodate the process used, or the process must be defined to meet structural design needs.

Castings are excellent candidates for three-dimensional complex shapes, and can be very cost effective in comparison to extrusions, forgings, and fabricated sheet and plate parts. Alloy A356 (same as 356 but has lower impurity levels) or 356 is used for many applications: examples are cast bases for lighting standards, aerospace castings, and automotive parts. Alloy A444-T4, given in Table 2.2, is a very ductile, relatively low strength alloy that has been used for bridge railing posts.

Forgings are employed when a high-quality three-dimensional shape is needed. They are more costly (price per unit weight) than other product forms because of costs of dies and equipment, and the number of alloys normally used in this process is smaller than for the other processes. Aerospace parts and truck and passenger car wheels are representative uses for this product form.

Extrusions are an excellent choice for parts with an intricate cross section that is constant along the length. "Soft alloys" 6063 and 6061 are the best choice for general-purpose use (e.g., structural shapes, tubes, architectural sections). The 5XXX alloys are harder to extrude, and thus are more expensive. They have been used, however, when high ductility is desired in the welded construction. "Hard-alloy" extrusions, such as 7075, are used for aerospace applications because of their high strength but are considerably higher in price than soft-alloy extrusions.

Sheet and plate products are available in most of the alloys made. These products have wide usage in buildings, containers, and tanks. Alloy types 3XXX, 5XXX, and 6XXX are typically used for general applications. The 2XXX and 7XXX alloys are generally employed in the aerospace market. The sheet and plate products can be formed into shapes and sections that are joined together to achieve complex shapes, or they can be machined to final shape.

Fabrication method. The product forms usually are cut or machined, perhaps bent to shape and joined together in order to achieve the final product. In general, aluminum alloys are readily fabricated, but ease of fabrication does vary.

High formability is important in some of the automotive applications. Alloys 5182-0, 6009-T4, and 2008-T4 are representative choices for this application. Other alloys can be formed but will be more limited as to the shape possible.

For applications in which welding is required, the strain-hardened 5XXX alloys probably provide the best combination of as-welded strength and structural toughness. Heat-treatable alloys such as 6063 and 6061 are readily welded but generally have lower strength and ductility. Subsequent thermal treatment improves the strength of welded heat-treatable alloys in some cases, but it does not improve ductility.

All the alloys can be cut, machined, and drilled. Mechanical fastening and adhesive bonding are feasible with all the alloys.

Environment.   Detailed descriptions of the various types of corrosion that aluminum alloys are subject to have been given previously.[4] General corrosion resistance of 3XXX, 5XXX, and 6XXX alloys in industrial and seacoast environments is better than that for the 2XXX and 7XXX alloys. Alcladding is sometimes used to improve corrosion resistance but it adds cost. The 3XXX, 5XXX, and 6XXX alloys also are generally not as susceptible to stress-corrosion cracking as the 2XXX and 7XXX alloys.

Temperature exposure can change both mechanical properties and corrosion resistance. At temperatures below room temperature, the various properties of the aluminum alloys steadily improve as the temperature decreases. At elevated temperatures, mechanical strength properties are decreased. Also, the corrosion resistance of some 5XXX alloys (high-magnesium alloys such as 5083 and 5456) is reduced on exposure to sustained elevated temperature because of metallurgical changes. Alloys with low magnesium—i.e., 5454 and 5052—are not affected. The corrosion resistance of other alloys is not affected by temperature.

Joints usually are the regions of the structure most susceptible to fatigue cracks caused by repetitive loads. Thus, product forms such as extrusions, forgings, and castings may be designed to eliminate joints in critical stress areas.

Maintainability.   Most of the general-purpose applications use unpainted 3XXX, 5XXX, and 6XXX alloys for long-term corrosion resistance without maintenance. Also, structures of these alloys may be repaired by all of the joining processes: welding, adhesive bonding, and mechanical fastening.

Cost.   The selling price of a product is related to the cost of making it. However, ingot is traded as a commodity and *its* price varies considerably. (Some information is given in Chap. 13.) Therefore, the information on product prices presented here is expressed in relative terms. Differences in the price of the product depend on alloy, product type, configuration, and quantity of parts needed. Figure 2.13 provides some general observations on prices for various product forms and alloys for preliminary design. Prices for final design should be obtained from the supplier. The relative prices per unit weight are used. The information presented is for material only; fabrication costs and effects of design changes because of strength or quality differences are not included.

Sheet and plate products of the lower-strength, non-heat-treatable alloys are the lowest-priced. Moderate-strength alloys requiring thermal treatments are of moderate price. The price tends to be higher as the alloy strength increases. For extrusions, solid sections of 6XXX alloys have the lowest price per unit weight. The price increases for semihollow shapes and is highest for hollow shapes. The price also increases with size of extrusion (press size needed). Other alloys, such as the 5XXX series, are more difficult to extrude, and thus are higher in price. The price of forgings depend on the type of forging and increases with complexity and size of the part and with the strength of the alloy. The price of castings is influenced by type of casting, number of parts produced, size and complexity of the part, and alloy. The final selection should trade off alloy performance versus alloy price.

Generally, the cost of a three-dimensional part, cast, forged, or extruded to final shape, will be economical compared to a similar part

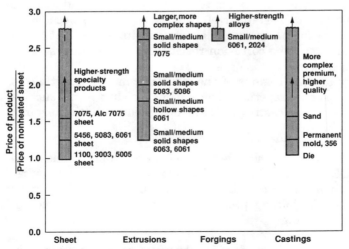

Figure 2.13   Relative price of aluminum products.

fabricated from less expensive sheet. Note that for each product form there are large differences in price depending on the difficulty in manufacturing, the alloy, and the number of parts produced. The price of the product thus depends on many factors, including ingot price. Generally, the lower-priced sheet and extrusions cost 25 to 50 percent more than the ingot price.

2.5   Mechanical Properties of Alloys

Table 2.3 provides "minimum" mechanical properties for several of the common general-purpose alloys. The minimum tensile properties are established from producer-generated data and are set so that 99 percent of the material is expected to conform at a confidence level of 0.95.[5] The other values are minimum expected properties developed from more limited test data. Most of the properties are from tests described by the appropriate ASTM standards. The tensile properties of castings are from separately cast test bars.[6] The minimum properties are normally employed for design.

Tensile stress-strain curves are provided for a few alloys in Fig. 2.14. Although there is an initial linear elastic portion of the curve, there is not a well-defined yield strength. The yield strength values for all alloys are defined as the stress corresponding to a 0.2 percent offset of strain from the initial straight-line portion of the curve.

Figures 2.15 to 2.18 show examples of the variation of mechanical properties with temperature,[7] based on characteristics at room temperature (24°C). Noteworthy is that all properties increase with decrease in temperature below room temperature. Aluminum alloys do not show a transition temperature as do most steels, and thus are good choices for applications involving low temperatures. The strength drops off at high temperatures. Creep needs to be considered in high-temperature use. Elongation remains high for all temperatures, low and high.

Table 2.4 shows representative strengths of two alloys under sustained load and elevated temperature as a fraction of room-temperature strength. The strength decreases as the time under stress increases. Alloy 6061 shows more resistance to creep (higher strengths) than 5086. Little creep occurs at room temperature unless stresses are near the tensile strength of the alloy.

The mechanical properties of aluminum alloys increase with increase in strain rate. Figure 2.19 shows some trends based on tests of various alloys reported in the literature. For most structural applications, it is reasonable and conservative to use test data for quasi-static loading for problems or applications with higher strain rates.

TABLE 2.3 Minimum Mechanical Properties of General-Purpose Alloys

Alloy and temper	Product	Thickness, in (mm)	Tension Ultimate, ksi (MPa)	Tension Yield, ksi (MPa)	Compression Yield, ksi (MPa)	Shear Ultimate, ksi (MPa)	Shear Yield, ksi (MPa)	Bearing Ultimate, ksi (MPa)	Bearing Yield, ksi (MPa)	Compressive modulus, ksi (MPa × 10^{-4})	Elongation, % in 2 in (51 mm)
1100-H14	Sheet, plate, rolled rod, and bar	0.114–1.0 (2.9–25.0)	16 (110)	14 (95)	13 (90)	10 (70)	8 (55)	32 (220)	21 (145)	10100 (7.0)	6
Alclad 3004-H16	Sheet	0.057–0.162 (1.3–4.1)	35 (240)	30 (210)	28 (195)	20 (140)	17 (115)	66 (455)	45 (310)	10100 (7.0)	4
5052-H34	Sheet, plate, rolled rod, and bar	0.25–1.0 (6.3–25.0)	34 (235)	26 (180)	24 (165)	20 (140)	15 (105)	65 (450)	44 (305)	10200 (7.0)	10
5083-H321	Sheet and plate	0.188–1.5 (4.8–38.1)	44 (305)	31 (215)	26 (180)	26 (180)	18 (125)	84 (580)	53 (365)	10400 (7.2)	12
5086-H116	Sheet and plate	0.063–2.0 (1.6–50.0)	40 (275)	28 (195)	26 (180)	24 (165)	16 (110)	78 (540)	48 (330)	10400 (7.2)	8
5454-H34	Sheet and plate	0.25–1.0 (6.3–25.0)	39 (270)	29 (200)	27 (185)	23 (160)	17 (115)	74 (510)	49 (340)	10400 (7.2)	10
5456-H116	Sheet and plate	0.063–1.25 (1.6–31.8)	46 (320)	33 (230)	27 (185)	27 (185)	19 (130)	87 (600)	56 (385)	10400 (7.2)	10
6061-T6	Extrusion	Up through 1.0 (Up through 25.0)	38 (260)	35 (240)	35 (240)	24 (165)	20 (140)	80 (550)	56 (385)	10100 (7.0)	8
6061-T62	Sheet and plate	0.25–1.0 (6.3–25.0)	42 (290)	35 (240)	35 (240)	27 (185)	20 (140)	88 (605)	58 (400)	10100 (7.0)	9
6063-T5	Extrusion	Up through 0.50 (Up through 12.5)	22 (150)	16 (110)	16 (110)	13 (90)	9 (60)	46 (320)	26 (180)	10100 (7.0)	8
356-T6	Permanent mold	—	33 (230)	22 (150)	22 (150)	20 (140)	13 (90)	69 (475)	36 (250)	10400 (7.2)	3
A444-T4*	Permanent mold	—	20 (140)	7 (50)	7 (50)	12 (85)	4 (30)	38 (260)	11 (75)	10100 (7.0)	20

*Yield strengths based on limited data. Values for shear and bearing estimated from tensile properties.

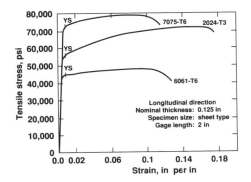

Figure 2.14 Representative stress-strain curves.

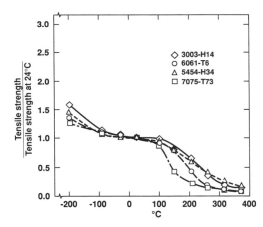

Figure 2.15 Effect of temperature on tensile strength.[7]

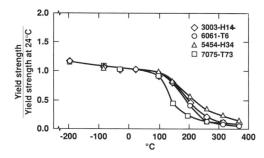

Figure 2.16 Effect of temperature on yield strength.[7]

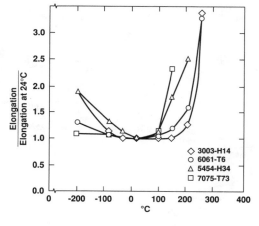

Figure 2.17 Effect of temperature on elongation.[7]

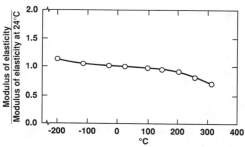

Figure 2.18 Effect of temperature on modulus of elasticity.[7]

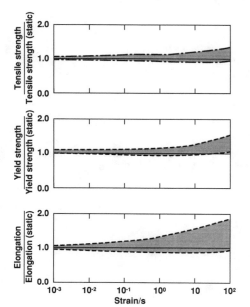

Figure 2.19 Properties under high strain rates.

TABLE 2.4 Creep Properties of Two Alloys*

Temperature, °F (°C)	Time under stress, h	6061-T6					5086-H34				
		Rupture	Creep, %				Rupture	Creep, %			
			1.0	0.5	0.2	0.1		1.0	0.5	0.2	0.1
75 (24)	0.1	1.0	1.0	0.98	0.96	0.93	1.0	1.0	1.0	1.0	—
	1	1.0	1.0	0.96	0.93	0.93	1.0	1.0	1.0	0.98	—
	10	1.0	0.98	0.96	0.93	0.93	1.0	1.0	1.0	0.85	—
	100	1.0	0.98	0.93	0.93	0.91	1.0	0.98	0.94	—	—
	1000	1.0	0.96	0.93	0.91	0.91	1.0	—	—	—	—
212 (100)	0.1	0.91	0.89	0.89	0.87	0.84	0.91	0.83	0.77	0.70	0.68
	1	0.89	0.87	0.87	0.84	0.84	0.85	0.77	0.72	0.68	0.60
	10	0.87	0.84	0.84	0.84	0.82	0.79	0.70	0.66	0.49	0.34
	100	0.84	0.84	0.84	0.82	0.82	0.68	0.55	0.43	0.30	0.21
	1000	0.82	0.82	0.82	0.82	0.80	0.57	0.34	0.30	0.20	—
300 (149)	0.1	0.82	0.80	0.80	0.78	0.78	0.79	0.70	0.66	0.55	0.43
	1	0.80	0.78	0.78	0.78	0.76	0.70	0.55	0.45	0.30	0.20
	10	0.78	0.76	0.76	0.76	0.73	0.57	0.32	0.26	0.18	—
	100	0.76	0.76	0.76	0.73	0.71	0.43	0.20	0.18	—	—
	1000	0.69	0.69	0.69	0.69	0.67	0.30	—	—	—	—
400 (204)	0.1	0.71	0.71	0.71	0.69	0.67	0.55	0.30	0.21	0.16	0.11
	1	0.67	0.67	0.67	0.64	0.62	0.43	0.17	0.14	0.09	0.06
	10	0.60	0.60	0.58	0.56	0.53	0.30	0.12	0.09	0.06	0.03
	100	0.51	0.49	0.49	0.47	0.42	0.19	0.08	0.06	0.03	—
	1000	0.40	0.40	0.40	0.36	0.31	0.14	0.05	0.04	0.02	—

*Figures are ratio of tensile strength at various temperatures and times under stress to tensile strength at 75°F (24°C).

References

1. Van Horn, Kent R. (ed.), *Aluminum,* Vol. 2, "Design and Application," American Society for Metals, Metals Park, Ohio, 1967.
2. "Historical Supplement," *Aluminum Statistical Review,* The Aluminum Association, Washington, D.C., 1982.
3. Staley, J. T., "Research on High-Strength Aerospace Aluminum Alloys," *Canadian Aeronautics and Space Journal,* Vol. 31, No. 1, March 1985.
4. Van Horn, Kent R. (ed.), *Aluminum,* Vol. 1, "Properties, Physical Metallurgy and Phase Diagrams," American Society for Metals, Metals Park, Ohio, 1967.
5. "Specifications for Aluminum Structures," *Aluminum Construction Manual,* Sec. 1, The Aluminum Association, Washington, D.C., 1986.
6. *Standards for Aluminum Sand and Permanent Mold Castings,* The Aluminum Association, Washington, D.C., March 1980.
7. "Engineering Data for Aluminum Structures," *Aluminum Construction Manual,* Sec. 3, The Aluminum Association, Washington, D.C., Nov. 1981.

3

Design Considerations

The emphasis of this book is on the structural design of an aluminum product. Structural engineering is one of several disciplines that normally would form a design team for the development of a complex product. Table 3.1 provides an example of such a team, with disciplines and responsibilities. The responsibilities can be combined in one or more knowledgeable, experienced designers for the development of simpler products.

TABLE 3.1 Design Team for Aluminum Product

Discipline	Responsibility
System behavior	Aesthetics, functional needs
Structural	Strength, stiffness, durability of components, joints, and system
Materials (alloy design)	Material characteristics in environment
Corrosion	Corrosive effects of environment on alloys and joints
Manufacturing	Practices for processes; joining, forming, cutting, machining, and assembly technology

The design of a complex product may take years of development with a large staff of people. Books are devoted to the design process.[1,2] The treatment here is brief. The discussion is included because it is important that the structural designer be aware of all the requirements for the product so that the design meets all needs. Figure 3.1

Figure 3.1 The design process.

provides the format for discussion: five important design steps that lead to the implementation of a product.

3.1 Environment

3.1.1 End use/function

The end use for the product usually is obvious and known from the start of development. However, alloy selection and design of the product are quite different depending on the application. A designer should not design a part, even a simple beam, without knowing what it will be used for, how it is connected to the final system, and how it will have to function in the final system. The following example illustrates the difficulty of defining "environment."

Figure 3.2 shows cracks that developed in the hull of the Alcoa *Seaprobe* after a few years of service. The cracks are at the tip of a tripping bracket, a logical location for fatigue cracks. The unusual feature was that only two cracks developed, one on the starboard and one on the port side in the same location in the forward portion of the ship. Hundreds of other identical details showed no problem, and a finite element analysis indicated low stresses in the hull plate, and thus no reason for the failures. A careful inspection of the area, however, showed that, at these two locations, some light auxiliary framing had been installed to support fireproofing panels (see Fig. 3.3). Analysis showed that the restraint imposed by these auxiliary members was

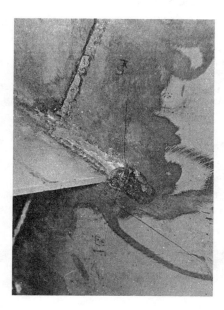

Figure 3.2   Cracks in hull at edge of tripping bracket.

Figure 3.3 Auxiliary framing for fireproofing.

sufficient to create a "hard spot" and high stresses in the hull at these two locations. The slamming forces on the hull in the forward area provided the fatigue mechanism for the cracking. The framing for the fireproofing was different in the rest of the ship and did not cause a problem. The restraining influence of these added members for fireproofing should have been but obviously was not considered by the structural engineer in the original hull design.

3.1.2 Loadings

Loads are sometimes not known accurately and the design process may have to include some field or laboratory measurements to determine realistic values. Included are mechanical loads such as truckloads on bridges, inertia loads on automotive frames, wind pressure (or vacuum) on roofing or siding products, and pressure (or vacuum) in beverage cans and pressure vessels. Thermally induced loads from nonuniform temperature in the aluminum structure or from a temperature change in structures that are a combination of aluminum attached to other materials (e.g., concrete or steel) with different coefficients of thermal expansion also need to be considered.

Loads on the elements, and consequently, stresses, are also introduced when products are fabricated and assembled. Even carefully fabricated parts have some inaccuracies in geometry, so they do not fit together without straining the parts. Figure 3.4 shows a mismatch causing stresses to be introduced in the chords of an overhead sign

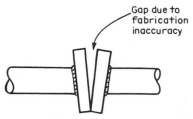

Figure 3.4 Mismatch of bolting flanges in tubular truss.

structure. The bolting flanges were not perfectly aligned. Two problems can result in such cases. First, too large a mismatch can cause
damage to the material when the parts are drawn together. Second, if
the parts are not drawn tightly together, the bolts will be loaded and
unloaded as the truss is loaded and unloaded. For cases in which cyclic
loading is present, fatigue failure of the bolts occurs. This problem is
much less severe if the parts are tight together and the bolts are under
prestress.

The flow of wind against and around structures not only creates
pressure and vacuum loadings but also can cause vibration of the
member or the entire structure. Figure 3.5 shows a failure when the
entire overhead sign structure vibrated. The failure in Fig. 3.6 is of a
tube in a truss, also caused by wind-induced vibration. The member
vibrated normal to the direction of the wind, and the appearance of
the failure is different from those when the entire structure vibrated.
The periodic transverse forces are the result of vortices shedding from
the members. Most structures can vibrate in the wind under appropriate conditions; the best design philosophy is to prevent the vibration.

Loads may be applied slowly (creep) or quickly (impact). The loads
may be relatively constant over the life of the product or may be cyclic.
All these conditions require different kinds of designs for the product.

There are some codes and standards that specify the loads that must
be used. Building and bridge design[3,4] have provided some guidance.
Subsequent chapters of this book will deal in more depth with design

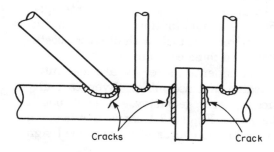

Figure 3.5 Cracks for overall vibration of overhead sign truss.

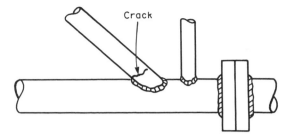

Figure 3.6   Cracks from vibration of individual member.

for wind-induced vibration and with the various codes and standards for aluminum that specify requirements for design of aluminum products and structures.

3.1.3   Atmospheric/other related environments

Elevated and low temperatures affect mechanical properties of aluminum alloys. Sustained elevated temperature can degrade the corrosion resistance of the 5XXX alloys with high magnesium levels.[5] Although most alloys have excellent resistance to industrial and seacoast atmospheres, some conditions can be very detrimental to aluminum.[5] Some acid and alkaline solutions, at elevated temperature, etched away the metal in an aluminum chimney to the point that it could no longer carry the wind loads. The mechanism was similar to that employed in the chemical milling of aluminum shapes.[6]

3.2   Performance Requirements

Aesthetics may be important. Size, shape, repairability, and replaceability of parts may be specified. The owner and/or user must be satisfied, as well as the designer developing the product.

Durability.   Existing aluminum products meet a wide variety of requirements in terms of life span, and these obviously influence design. Beverage cans have a life of a few months, so the ability to recycle is important, not long life. Automobiles, with lives of about 10 years or less, building sheathing (20 years or more); aircraft (over 30 years); and infrastructures and major bridge structures (normally maintained to 80–100 years or longer) are examples of long-life needs. Design information is available for aluminum to allow for the design life required, although long-term durability of any product of any material is difficult to predict without experience in the application. Fatigue, corrosion, and combined effects are the primary ingredients to be considered in designing for durability.

Maintainability.   All of the long-life applications should be relatively free of maintenance. The choice of aluminum for many current uses has been made in part because the product is relatively maintenance-free. Considerations in this category are the need for inspections, cleaning methods, and repair techniques.

Safety.   There are almost always uncertainties in the design of a product. Loads usually are not precisely known. Dead loads may be accurately calculated for a particular design, but live loads, such as wind loads, vehicle loads on bridges, and inertia loads for a vehicle frame, are less certain.

Other variables are material properties, although generally statistical minimums are available for aluminum alloys. The as-fabricated geometry is not perfect; the size of parts varies somewhat and imperfections such as crookedness of parts and fabrication defects are introduced. The strength of most parts can be estimated, but not precisely. All these factors introduce uncertainty into the design. The uncertainties are taken into account in specifications by the use of a factor of safety or load and resistance factors. The Aluminum Association specifications[7] give explicit factors to be applied to the calculated yield and ultimate strength of the components; this method is called *allowable stress design*. The factors of safety employed in these specifications are given in Table 3.2. A factor is applied to ultimate strength (collapse) and to yielding (serviceability). Note that the factors are more conservative for bridges than buildings. Considerations of consequences of structural failure on loss of life and property are implicit in these factors.

An alternate procedure to allowable stress design is to employ factors that are dependent on the uncertainty of the load and behavior of the component. This design procedure is referred to as *load and resistance factor design (LRFD)*. Much of the work to develop this design procedure has been done, and U.S. codes are being developed accord-

TABLE 3.2   Factors of Safety In The Aluminum Association Specifications

Components	Buildings	Bridges
Columns	1.95 ultimate 1.65 yielding	2.2 ultimate 1.85 yielding
Tension members	1.95 ultimate 1.65 yielding	2.2 ultimate 1.85 yielding
Connections	2.34 ultimate	2.64 ultimate

ingly. Canadian[8] and European design guidelines utilize this method. An example of load factors is as follows:[9]

$$F = 1.2D + 1.6L \qquad (3.1)$$

where F = factored load
    D = dead load
    L = live load

The factored load is represented by a sum of the important types of load. A smaller multiplier is applied to the dead load because it can be accurately calculated, and a larger multiplier is applied to live load, which is less precisely known. The stresses in the part calculated from the factored loads must be less than the resistance (strength) of the part:

$$R = \phi S \qquad (3.2)$$

where R = resistance of the part
    S = calculated strength of the part
    ϕ = factor that accounts for all uncertainties in the calculated
        strength of the part (generally 0.8 to 0.9)

For the part to be designed safely the resistance of the part must be equal to or greater than the imposed loading:

$$R \geq F \qquad (3.3)$$

The primary purpose of this book is to provide information concerning the strength design of aluminum structures. Subsequent chapters are devoted to this subject in detail. In applying the information, the designer must employ a "factor of safety" commensurate with the uncertainties of load and resistance, and with the importance of the application.[10]

3.2.1 Cost

The cost of aluminum products has several ingredients; each or all may be important depending on the application.

- Cost of the material
- Cost of fabricating the material into assemblies
- Cost of assembling or erecting the final product
- Cost of maintaining or operating the product
- Cost of disposing of the product after its useful life

Although the initial cost of the product is important, a more reasonable and increasingly popular view is to consider the life-cycle cost of the product. The life-cycle cost is the total of the five costs tabulated above.

The cost (price) of the material was discussed in Chap. 2 and is dependent on the alloy, complexity of the part, type of mill product, and amount of the mill product purchased. The other costs are discussed briefly in the remaining chapters but vary substantially depending on the nature of the final products. Cost figures are best supplied by companies that do the work. Fabrication of the metal into components requires processes such as forming, joining, machining, and finishing. Assembly or erection joins components and assemblies of components into the final product form. Sometimes the light weight of the aluminum parts is an advantage in assembly and erection.

Maintenance may include cleaning, painting, and repairing. Operating costs for transportation vehicles are usually associated with fuel consumption. Disposal of used aluminum products is usually easy because there is an excellent market for recycled aluminum.

3.3 Concept

After the environment and the performance requirements are known, the designer is in a position to establish one or more conceptual designs. Geometry of the structure is set based on the application needs and available experience. The aluminum alloys and tempers and the product forms are selected. Most products are assemblies of components, so decisions must be made about fabrication processes, joining methods, etc. Preliminary plans for manufacturing or erection of the product are also needed, so that an estimate of the final cost of the product can be established. An evaluation should also be made of the amount of effort needed to complete the design of the product.

The amount of effort expended in conceptual design is small compared to that needed for a detailed design, and the design tools may be relatively simple, such as rules-of-thumb and hand calculations. The largest effort is needed for final design to evaluate and optimize the selected concept.

3.4 Evaluation and Optimization

This step begins with the selected concept and optimizes it based on weight, cost, or other specified characteristics. Generally, the design is not sufficient if the product simply meets the requirements; the design must be optimal, otherwise it will not be or will not continue to be competitive in the marketplace. The behavior of the aluminum structures covered in later chapters can be a key guide for structural design.

Global behavior can often be estimated by the use of finite element simulations. A number of general-purpose codes are available. Some are given in Table 3.3 with comments on their usefulness for various kinds of structural simulations.[11-16] Components are generally sized using engineering solutions of the type given in subsequent chapters. In the optimization of the product the designer may make use of classical optimization techniques and computer codes, simpler hand calculations, or trial-and-error techniques.

Although it is economical to use analytical techniques as much as possible in final design, some experimental evaluation of components and of the final concept is essential. It is nearly impossible for the designer to anticipate all problems unless the product is very similar to an existing satisfactory design. Other disciplines such as manufacturing will be developing the technology needed for the product, parallel to but in cooperation with the work being done by the structural engineer.

Plans for final manufacturing or erection of the product are developed. Variability of the geometry of basic products, components, and assembly must be considered. The design must be robust so that the variations in material properties and geometry will not produce an excessive number of products that do not meet the requirements. When all possible problems have been evaluated by the designer, the product is ready for the customer; the manufacture can begin.

TABLE 3.3   General-Purpose Computer Codes

Code	Technical strengths of code	Primary uses	Usefulness for buckling and crippling analyses (thin sections)
ANSYS	User oriented Large element library Graphics	Elastic stress analysis Small displacement analyses Modal and dynamic analyses	Not used
ABAQUS	Nonlinear structural/ heat transfer	Nonlinear structural analyses Process simulations	Axisymmetric shell behavior Laminated shells
DYNA3D	Large deformation/ dynamic Contact algorithms	Buckling and crippling Impact	All quasi-static and dynamic
NIKE2D	Large strains and rotations	Sheet forming	Under evaluation

References

1. Wallace, Ken (ed.), *G. Pahl–W. Beitz Engineering Design* (first English version), The Design Council, London, 1984.
2. French, Michael J., *Conceptual Design for Engineers,* 2d ed., The Design Council, London; Springer-Verlag, Berlin, Heidelberg, New York, Tokyo, 1985.
3. *Manual of Steel Construction,* American Institute of Steel Construction, New York, 1970.
4. AASHTO, *Standard Specifications for Highway Bridges,* 13th ed., American Association of State Highway and Transportation Officials, Washington, D.C., 1983.
5. Van Horn, Kent R. (ed.), *Aluminum,* Vol. 1, "Properties, Physical Metallurgy and Phase Diagrams," American Society for Metals, Metals Park, Ohio, 1967.
6. Van Horn, Kent R. (ed.), *Aluminum,* Vol. 3, "Fabrication and Finishing," American Society for Metals, Metals Park, Ohio, 1967.
7. "Specifications for Aluminum Structures," *Aluminum Construction Manual,* Sec. 1, The Aluminum Association, Washington, D.C., 1986.
8. Marsh, Cedric, *Strength of Aluminum,* 5th ed., The Aluminum Company of Canada Limited, Montreal, Canada, 1983.
9. American National Standard, *Minimum Design Loads for Buildings and Other Structures,* ANSI A58.1, 1982.
10. Hinkle, A. J., and Sharp, M. L., "Load Resistance Factor Design of Aluminum Structures," *Structural Safety and Reliability,* Vol. 3, (A. H-S. Ang, M. Shinozuka, and G. I. Schuëller, eds.), American Society of Civil Engineers, New York, *Proceedings of ICOSSAR '89,* The 5th International Conference on Structural Safety and Reliability, San Francisco, California, August 7–11, 1989.
11. DeSalvo, Bariel J., and Gorman, Robert W., ANSYS Engineering Analysis System, User's Manual, Vol. 1 and Vol. 2, Rev. 4.3, Swanson Analysis Systems, Inc., Houston, Pennsylvania, June 1, 1987.
12. Hibbitt, Karlsson & Sorenson, Inc., ABAQUS User's Manual, Rev. 4.7, Providence, Rhode Island, 1988.
13. Hallquist, J. O., and Benson, D. J., "DYNA3D User's Manual—Nonlinear Dynamic Analysis of Structures in Three Dimensions," University of California, Lawrence Livermore National Laboratory, Rept. UCID-19592, Rev. 2, March 1986.
14. Hallquist, J. O., "NIKE3D: An Implicit, Finite Deformation, Finite Element Code for Analyzing the Static and Dynamic Response of Three-Dimensional Solids," University of California, Lawrence Livermore National Laboratory, Rept. UCID-18822, January 1981.
15. Marc Analysis Research Corporation, *MARC General Purpose Finite Element Program,* Vol. C, "Program Input," Palo Alto, California, 1983.
16. Sharp, M. L., Dick, R. E., Banthia, V. K., and Kulak, M., "Predictions of Buckling/ Crippling Behavior of Thin Plates and Shells," *Structural Stability Research Council Annual Technical Session Proceedings,* Annual Technical Meeting, St. Louis, Missouri, April 10–11, 1990.

Chapter

4

Structural Design

The spaceframe concept for an aluminum automotive frame is illustrated in Fig. 4.1.[1] This structure will be used to define some of the terminology used herein as well as to discuss various failure modes to be considered by the designer. The geometry of the spaceframe, location of members, shape and size of the members, location and magnitude of masses, location and type of supports at the suspension, and type of joints are established by the designer based on design requirements, experience, and calculations for local load requirements. The structural frame must be consistent with the intended styling and packaging (space for people, luggage, machinery, and so forth), and thus has many restrictions on size and shape of elements. The concept

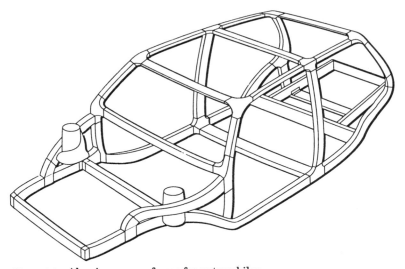

Figure 4.1 Aluminum spaceframe for automobiles.

shown is a latticed framework of beams and columns joined together at their ends. Overhead sign structures and roof spaceframes are similar examples to the concept shown. Other common aluminum applications such as aircraft structures, boats and ships, and storage tanks are constructed of stiffened sheet or plate rather than beams and columns. Some of the failure modes for this type of construction are also discussed in this chapter.

In order that the automotive frame perform as needed for the application, various characteristics (examples are shown in Fig. 4.2) must be addressed. Some of these considerations require an analysis of the entire structure (global structure). Other requirements may be resolved by a component (member) or element analysis. Stiffness and vibration frequency, important for handling quality, are obtained from an analysis of the entire frame. Energy absorption, strength, and toughness of joints depend on the alloy and the proportions of the components and the type and design of the joints. The design of the joint is also critical for fatigue strength. If fatigue cracks initiate, most will do so at these locations. Both global and component analyses are needed to evaluate corrosion resistance.

The smallest subdivisions of the structure to consider are the elements of the component, generally the flat or curved parts making up the shape. Failure of the elements usually will seriously reduce the strength of the component and global structure, and thus must be avoided.

The structural designer thus needs to consider possible failure modes for global structures, their components, elements of the components, and any combined interaction, such as the influence of element failure on component failure.

This chapter provides information on some of the failure modes that must be considered. Subsequent chapters will provide specific guidelines for analyzing components and elements, and the interaction among failure modes. A few global structural problems will be addressed, but, in general, the designer will need to make use of avail-

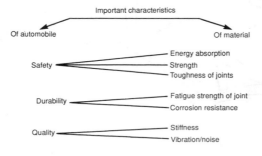

Figure 4.2 Structural considerations—automotive frame.

able computer codes and other analysis tools for the global response. The design guidelines will generally estimate failure conditions; therefore, the designer will need to apply appropriate safety margins based on the uncertainty of the loads and member strength. The margin of safety can be applied by increasing the applied loads and comparing these increased loads to member strengths, or by reducing the design stress of the member and comparing this reduced stress to the stress from applied loads.

4.1 Global Structure

The automotive structure shown in Fig. 4.1 is supported at the attachment points of the suspension and is subjected to the specified thermal and mechanical loads. Some of the failure conditions that the designer is concerned about are static and fatigue strength, excessive deflection, excessive deceleration in a crash, and too low a natural frequency of vibration. Some of the failure conditions result in fracture of the aluminum part, others result in a product that does not meet functional requirements. For example, a frame with excessive deformation (low stiffness) under load and low frequencies of vibration will not meet quality standards for handling and noise. Controlled collapse of the parts of the frame in a crash is a necessary safety feature: too fast a deceleration is fatal to the occupants. The passenger compartment is made strong enough to protect the occupants in a crash. Durability, the ability of the frame to function satisfactorily over the desired life under fatigue and corrosion, is required. Other failure modes may be possible depending on the structure. The designer needs to become as familiar as possible with prior experience so that potential failure modes are identified. For example, latticed structures exposed to the wind can vibrate, either as a global structure or as an individual member, resulting in fatigue failures. This problem is discussed in more detail in Chap. 12. Single-layer latticed aluminum structures and stiffened and unstiffened shells may fail by buckling. In most cases, the mechanical characteristics of the alloy will be important, but the method and accuracy of manufacturing can have equal or greater importance.

The behavior of the global structure generally is obtained from a mathematical simulation utilizing a finite element code. Linear-elastic solutions often are satisfactory for stress analysis or frequency-of-vibration calculations. Geometrical and material nonlinearities must be considered in crash modeling or to estimate other failure conditions. Often experiments are run to verify the accuracy of the calculations.

A source of potentially large errors in a global analysis is inaccuracy in describing the characteristics of the joints and supports. Joints used to attach members are rarely rigid, and the correct joint stiffnesses must be included in calculations; the assumption of rigid joints may result in large errors in calculated deflection and frequency of vibration. In addition, the joints need adequate strength so that the structure has good toughness. Behavior characteristics for the joints may be obtained by a careful theoretical analysis of each joint or by tests.

The rigidity and strength of the supports affect the way the structure carries its loads; this is particularly true of structures with redundant supports. Figure 4.3 shows a test setup for a model of a roof spaceframe of irregular plan shape, supported at three locations, with six reaction points at each location. The buckets hung from the model are to contain the lead shot used to obtain various loading cases. Figure 4.4 shows a close-up of one of the support areas. The support rods are instrumented so that individual reactions are obtained. Figure 4.5 shows the variation of one of the reactions as the stiffness of the support is changed. The reaction changes by a factor of about 5. Obviously, the loads in the members near the reactions also change considerably.

The global analysis provides the loading that components, elements, and joints need to sustain. Some of the various failure modes of these subelements are reviewed in the next section.

Figure 4.3 Test of model of latticed spaceframe.

Figure 4.4 Close-up of support area of spaceframe.

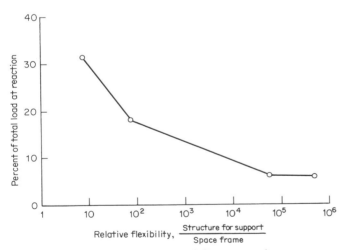

Figure 4.5 Influence of support flexibility on reactions.

4.2 Components

Figure 4.6a shows a component that could be part of a frame. Because of the ease of the fabrication process for aluminum extrusions, the material can be placed in the most advantageous location for the imposed loads; this results in good efficiency but shapes that are often complex in section. Usually, thin-walled sections will be employed; there is no need for a corrosion allowance. Thus, buckling failures need to be considered for many aluminum components and elements. For the component shown, several buckling modes are possible depending on the imposed load. If the load is predominantly compressive, the member is a column, and failure can occur by buckling by flexure, torsion, or combined flexure-torsion (see Fig. 4.7). Closed-section (tubular) components as shown have excellent torsional stiffness and usually do not buckle by torsion. Open sections such as angles, tees, and I-sections are susceptible to buckling by torsion or flexure-torsion. Stiffened panels can also suffer flexural buckling.

If the loading is predominantly bending, the beam can fail by excessive deflection or fracture of the tensile flange, and, particularly if it is of open section, can fail by lateral buckling (see Fig. 4.8). Combined loadings with axial, bending, and torsional contributions produce a complicated failure behavior in the component. Usually the component is considered separately as a beam and as a column and the combined behavior is calculated using an interaction equation.

If the bending or axial loading produces high tensile stresses, particularly at joints, failure due to fracture of the material or excessive yielding is considered. Joints and notches usually affect the tensile strength of components and are critical if fatigue is a factor.

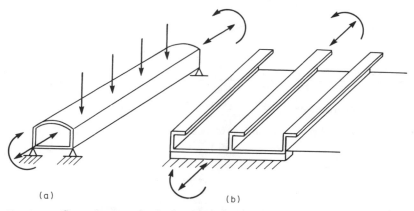

(a)

(b)

Figure 4.6 Components under load. (a) Beam, column, or tension member (part of a latticed structure). (b) Stiffened panel (part of a stiffened shell structure).

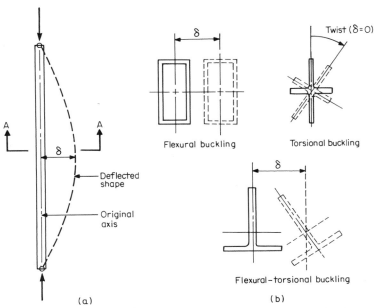

Figure 4.7 Failures in axially loaded columns. (*a*) Axially loaded columns. (*b*) Sections A-A showing deflections at failure.

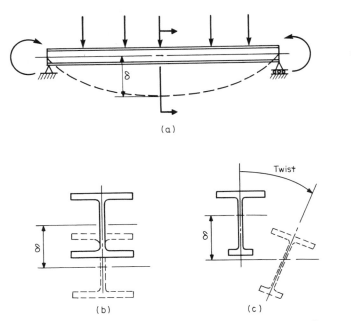

Figure 4.8 Failures in laterally loaded beams. (*a*) Laterally loaded beams. (*b*) Failure by excessive deflection or fracture. (*c*) Failure by lateral buckling.

One of the assumptions generally made in the design of components is that their ends are prevented from translation by the remaining components of the global structure. The conditions at all points of support are critical for calculating strength. Allowance must be made for any flexibility of the supports. The buckling load of components of the type mentioned is the highest load that they can sustain. There is no higher postbuckling strength of the type inherent in elements such as plates that are supported at all of their edges. Also, imperfections introduced during fabrication, such as an initial lateral bow, affect strength and must be considered in design.

4.3 Elements

Thin elements, either flat or curved, under compressive or shear loading, can buckle. Included are those elements with a stress gradient that can vary from tension to compression (such as the web of a beam under bending loads). These elements can be part of a stiffened panel as shown in Fig. 4.6b or part of a beam or column. Buckled shapes of some elements are illustrated on Fig. 4.9. The buckling load is affected by geometrical imperfections, type and distribution of stress, and support conditions. However, for most thin plates, supported on all edges, there is considerable strength above that calculated for elastic buckling and the ultimate strength, or "crippling" strength, defined in terms of the applied load that causes a failure condition. Note that if appearance of a part is important, the buckling load may be the maximum load allowed by the designer; the reason for this is that above that load a waviness becomes visible in the elements.

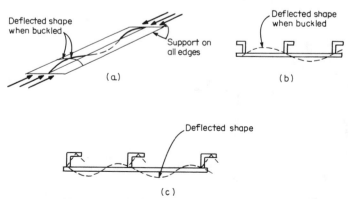

Figure 4.9 Shapes of buckled elements. (a) Plate in compression. (b) Stiffened plate in compression with buckling of elements. (c) Stiffened plate in compression with lateral buckling of stiffeners.

4.4 Joints

Joints include the attachments of elements together to form a component and the attachments of components to construct the total structure. Joints, unless carefully designed, can be the weak links in the structure; many service failures occur at connections. Joining is by mechanical methods such as bolts, rivets, and clinches; welding, both fusion and resistance; adhesive bonding; and combinations of these methods.

Fatigue failures usually occur at joints. Most joints have high localized stresses, compared to those existing in the elements or components. Under repetitive loading, cracks initiate at these high-stress locations. Residual stresses from welding also affect fatigue strength and buckling strength.

The tensile or compressive strength of most joints is less than that of the component. If the joint strength is less than the yield strength of the component, the structure can fail in a rather nonductile manner with little energy absorbed.

4.5 Structural Design

These initial chapters are intended to provide designers not familiar with aluminum an introduction to the metal and its characteristics. Also included are initial thoughts and experiences related to the behavior and design of aluminum structures. The remaining chapters will be devoted to a more detailed review of the behavior of elements, components, and global structures. The behavior is quantified, and available experimental and theoretical work is provided. It is believed that this work can be used with confidence by engineers involved in design of aluminum structures.

Reference

1. Winter, E. F. M., Sharp, M. L., Nordmark, G. E., and Banthia, V. K., "Design Considerations for Aluminum Spaceframe Automotive Structures," *Technical Papers*, Vol. II, 23d FISITA Congress, Torino, Italy, May 7–11, 1990.

5

Axially Loaded Members

5.1 Tension Components

For members loaded in static tension, there are two failure modes of interest, excessive plastic deformation and fracture. For the alloys and products considered in this book, sufficient ductility is present so that the design can be based on the average stress at the net section reaching either the yield strength or the tensile strength of the alloy. *Ductility* in this case refers to the capability of the alloy to accommodate locally high strains without fracture. The net section usually occurs at joints. The discussion in this part of the chapter deals with the ability of the alloys to accommodate static tensile loads. The design of joints for dynamic or fatigue behavior will be covered in a later chapter.

5.1.1 Effect of notches

The intentionally built in stress concentrations such as bolt holes and mild notches are normally neglected in static design. Figure 5.1 shows test data from edge-notched specimens of three alloys. Alloy 6061 is considered to be a ductile, general-purpose material and the other two alloys are higher-strength, somewhat less ductile materials normally used in aircraft structures. The tensile properties are given in Table 5.1. For specimens with stress concentrations corresponding to those expected for round holes or similar types of discontinuities, the failure stress on the net section equals or exceeds the tensile strength for all of the alloys. For the specimens with the sharp radii, 6061 still develops strengths exceeding the tensile strength; the strengths for speci-

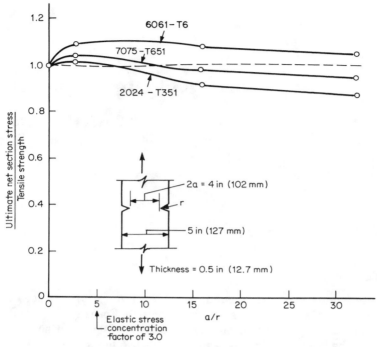

Figure 5.1   Effect of notches on tensile behavior.

TABLE 5.1   Tensile Properties of ½-in (12.7-mm) Plate*

	Tensile strength		Yield strength		Elongation in 2	Reduction
Alloy	ksi	MPa	ksi	MPa	in (50.8 mm), %	in area, %
2024-T351	70.0	483	53.6	370	22.7	27.4
6061-T651	46.8	323	44.6	308	19.7	49.1
7075-T651	85.5	590	78.6	542	14.5	24.0

*Material for specimens used in Fig. 5.1.

mens of the alloys are somewhat lower. Figure 5.2 shows a 50× magnification of the notches before and after failure. Large plastic deformation has occurred in each case. The deformation is approximately proportional to the reduction in area values given in Table 5.1. The local deformation allows the average stresses in the net section to reach the tensile strength of the alloy. A somewhat higher failure stress is achieved in these specimens because of the triaxial stress at the notch root.

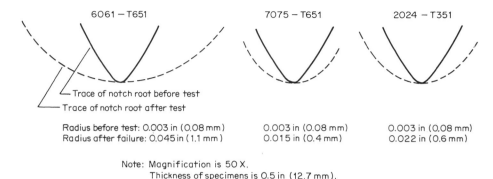

6061 – T651   7075 – T651   2024 – T351

Trace of notch root before test

Trace of notch root after test

Radius before test: 0.003 in (0.08 mm)   0.003 in (0.08 mm)   0.003 in (0.08 mm)
Radius after failure: 0.045 in (1.1 mm)   0.015 in (0.4 mm)   0.022 in (0.6 mm)

Note: Magnification is 50 X.
Thickness of specimens is 0.5 in (12.7 mm).

Figure 5.2   Deformation of notch at failure.

5.1.2 Effect of notches and localized loading

Angles attached by one leg by bolts must accommodate not only the local concentration at the holes but also the effects of the nonuniform loading on the shape. Figure 5.3 shows test data for single-angle and double-angle specimens. These tests provide a more severe measure of the ductility of an alloy. The ends of the single angles (one leg) were

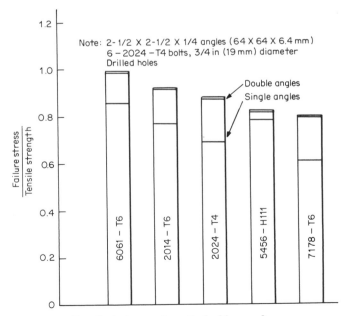

Note: 2-1/2 X 2-1/2 X 1/4 angles (64 X 64 X 6.4 mm)
6 – 2024 – T4 bolts, 3/4 in (19 mm) diameter
Drilled holes

Double angles
Single angles

$\dfrac{\text{Failure stress}}{\text{Tensile strength}}$

6061 – T6   2014 – T6   2024 – T4   5456 – H111   7178 – T6

Figure 5.3   Tensile tests—angles attached by one leg.

bolted to thin gusset plates, while one leg of each angle at each end of the double-angle specimens was bolted to a gusset plate. The double angles have less eccentricity of loading and, thus, higher strengths. Alloy 6061 specimens have average stresses close to the alloy's tensile strength, and alloy 7178 specimens have the lowest strength compared to the alloy's tensile strength. The design of members of this type must take into account the nonuniform loading; ductility cannot accommodate all of the redistribution of stresses needed to design based upon a nominal average stress. More design information is provided in Chap. 10 on joints.

A combination of low ductility and nonuniform loading can result in low load-carrying capacity of a part loaded in tension. Figure 5.4 shows the cross section of an aluminum part at the location at which it is attached to other members by bolts. The fabricating process used to make the part produced a reasonable tensile strength but an elongation of less than 1 percent (much lower than that acceptable for general structural materials). The net section stress at failure is about one-third of the tensile strength because the material could not accommodate the combined stress concentration and the nonuniform loading. Even if the material in the web is neglected and the flanges are

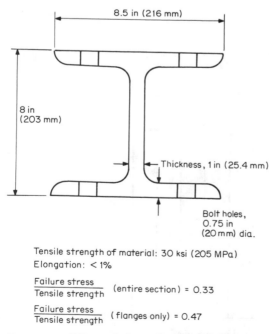

Tensile strength of material: 30 ksi (205 MPa)
Elongation: < 1%

$$\frac{\text{Failure stress}}{\text{Tensile strength}} \text{ (entire section)} = 0.33$$

$$\frac{\text{Failure stress}}{\text{Tensile strength}} \text{ (flanges only)} = 0.47$$

Figure 5.4 Effect of low ductility material on strength.

considered to carry the total load, the calculated average net section stress is less than one-half of the tensile strength. The behavior of parts made of low-ductility material is difficult to predict. Computer simulations or tests of parts are normally required for final design.

5.1.3 Effect of welds

Because most of the aluminum alloys achieve their strength by thermal treatment or strain-hardening, fusion welding causes a partial annealing of the material in a narrow area in the vicinity of the weld as illustrated in Fig. 5.5. The width of the reduced-strength zone is process dependent but for general design it is assumed to lie within 1.0 in (25 mm) of a weld. The amount by which strength is reduced is dependent on alloy. Table 5.2 gives properties for welded construction. The tensile strength given is 90 percent of the ASME weld qualification value. The yield strength is based on a 10-in (254-mm) gage length across a butt weld. The yield strength and the elongation always vary with gage length, but particularly so in this case because of the differences in properties across the butt weld. The load-deformation values using a 10-in (254-mm) gage length have been shown to be consistent with those for riveted and bolted joints.[2] The yield strength of the reduced-strength zone material as determined by standard tensile tests (specimens taken parallel to the weld) is about 75 percent of that given in Table 5.2 based on about 30 tests of butt-welded specimens of 6061-T6, 5083-H113, 5154-H34, 5454-H34, and 5456-H321.

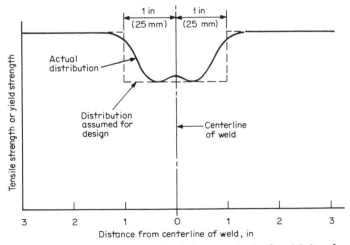

Figure 5.5 Distribution of mechanical properties in the vicinity of a weld.

TABLE 5.2 Minimum Mechanical Properties for Welded Aluminum Alloys[1]

Alloy and temper	Product	Thickness range, in (mm)	Tension		Compression	Shear		Bearing	
			Ultimate,* ksi (MPa)	Yield,† ksi (MPa)	Yield,† ksi (MPa)	Ultimate, ksi (MPa)	Yield, ksi (MPa)	Ultimate, ksi (MPa)	Yield, ksi (MPa)
1100-H12, H14	All	All	10 (70)	4.5 (30)	4.5 (30)	8 (55)	2.5 (15)	23 (160)	8 (55)
Alclad 3004-H16	All	All	19 (130)	11 (75)	11 (75)	13 (90)	6.5 (45)	44 (305)	19 (130)
5052-H34	All	All	22 (150)	13 (90)	13 (90)	16 (110)	7.5 (50)	50 (345)	19 (130)
5083-H321	Sheet and plate	0.188–1.500 (4.8–38)	36 (250)	24 (165)	24 (165)	24 (165)	14 (95)	80 (550)	36 (250)
5086-H116	Sheet and plate	All	32 (220)	19 (130)	19 (130)	21 (145)	11 (75)	70 (480)	28 (195)
5454-H34	Sheet and plate	All	28 (195)	16 (110)	16 (110)	19 (130)	9.5 (65)	62 (425)	24 (165)
5456-H116	Sheet and plate	0.188–1.500 (4.8–38)	38 (260)	26 (180)	24 (165)	25 (170)	15 (105)	84 (580)	38 (260)
6061-T6, T62‡	All	All‡	22 (150)	20 (140)	20 (140)	15 (105)	12 (85)	50 (345)	30 (210)
6061-T6, T62 (welded with 4043, 5554, 5654)	All	Over 0.375 (9.5)	22 (150)	15 (105)	15 (105)	15 (105)	9 (60)	50 (345)	30 (210)
6063-T5	All	All	15 (105)	11 (75)	11 (75)	11 (75)	6.5 (45)	34 (235)	22 (150)

*90% of ASME weld qualification test values.
†Corresponding to 0.2% offset in a 10-in (254-mm) gage length across a butt weld.
‡See values below for thicknesses over 0.375 in (9.5 mm) when filler alloys are 4043, 5554, or 5654.

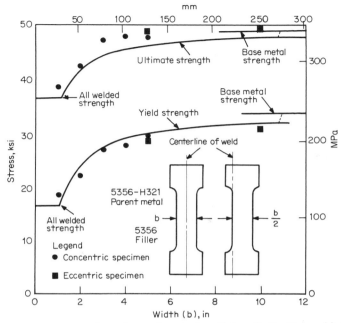

Figure 5.6 Results of tension tests of longitudinally welded members.[2]

Figure 5.6 shows that the tensile strength of specimens with longitudinal welds affecting part of the cross section can be predicted by the following equation:

$$\sigma = \sigma_b - (A_r/A)(\sigma_b - \sigma_r) \tag{5.1}$$

where σ = tensile or yield strength on longitudinally welded part
  σ_b = tensile or yield strength of base or unwelded material
  σ_r = tensile or yield strength of reduced-strength material
  A_r = area of reduced-strength material
  A = total area of components

The tensile strengths provided in Table 5.2 and yield strength values corresponding to 75 percent of the yield strengths given in Table 5.2 should be used in Eq. (5.1).

5.1.4 Effects of fatigue cracks

Designers of aircraft structures, in particular, investigate the potential detrimental effects of cracks that might develop in critical structures during service or defects that may be fabricated into the structure. The calculations are done using fracture mechanics techniques. Nonplane strain K_c or plane strain K_{Ic} values are defined for many of the high-strength aerospace alloys. Some values are given in Fig. 5.7.[3] This figure also shows that fracture toughness values vary approximately with the unit propagation energy from a tear test. The tear test is described in several papers[3,4] and the test has been used to characterize the behavior of most of the structural alloys. Some results are given in Table 5.3. Because the alloys used for general-purpose construction are too tough to obtain valid fracture toughness values there needs to be some way to estimate values. Figure 5.7 provides curves that can be used (extrapolated when necessary) so that the fracture toughness values can be estimated from the tear test results given in Table 5.3. These estimated values are believed to be useful for approximate studies. The calculations can determine either a critical size of crack or the strength of the part in the presence of a crack. The equation has the form

$$\sigma = \frac{K_{Ic}}{Y\sqrt{\pi a}} \quad \text{or} \quad \sigma = \frac{K_c}{Y\sqrt{\pi a}} \tag{5.2}$$

where σ = stress at fracture

K_{Ic}, K_c = plane strain, nonplane strain fracture toughness, respectively

a = half of crack length

Y = geometrical factor (assumed equal to 1.0 in Table 5.4)

The equations for a few cases are given in Table 5.4.[5]

5.1.5 Limiting proportions of tension components

Often members have secondary functions to perform and must be proportioned accordingly. For example, the component may need to support workers or equipment. The parts may need to be designed to prevent vibration from machinery or other sources of excitation that will cause fatigue failures (see Chap. 12). Most slender structural shapes will vibrate in the wind or in fluid flow, which also can lead to fatigue failures.

5.2 Compression Components

Columns may fail in a number of ways. If they are short enough, failure is by yielding and large plastic deformation will occur. Most prac-

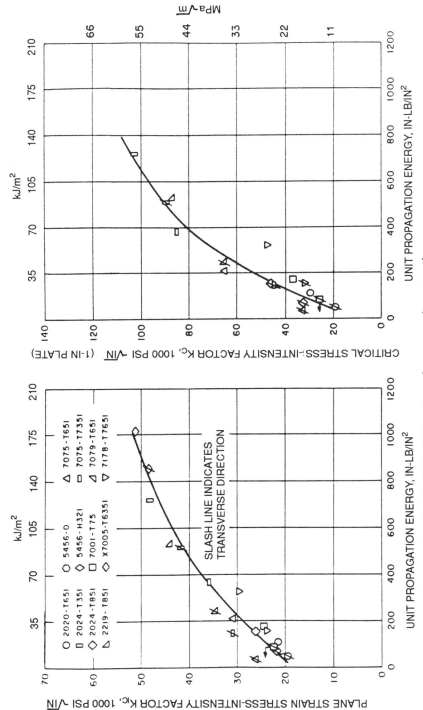

Figure 5.7 Fracture toughness values for 1-in-thick (25.4-mm) panels versus unit propagation energy from tear tests.

TABLE 5.3 Tear Test Data[*,3,4]

Alloy and temper	Longitudinal				Transverse			
	Tensile strength, ksi (MPa)	Yield strength, ksi (MPa)	Elongation, %	Unit propagation energy, in · lb/in² (kJ/m²)	Tensile strength, ksi (MPa)	Yield strength, ksi (MPa)	Elongation, %	Unit propagation energy, in · lb/in² (kJ/m²)
1100-H14	17.9 (123)	16.8 (116)	13.0	635 (111)	18.0 (124)	16.4 (113)	9.2	570 (100)
3004-H34	36.0 (248)	30.5 (210)	8.4	650 (114)	36.4 (251)	29.2 (201)	8.9	590 (103)
5052-H34	37.7 (260)	31.0 (214)	10.0	865 (151)	38.1 (263)	30.0 (207)	11.8	850 (149)
5086-H32	44.8 (309)	33.5 (231)	12.8	970 (170)	45.0 (310)	30.4 (210)	15.0	990 (173)
5456-0	47.5 (328)	22.6 (156)	20.8	930 (163)	48.3 (333)	22.8 (157)	22.2	935 (164)
6061-T6	45.5 (314)	41.5 (286)	12.4	900 (158)	45.2 (312)	39.8 (274)	12.5	740 (130)
7005-T7351	54.2 (374)	47.2 (325)	17.0	1015 (178)	53.3 (368)	46.5 (321)	16.2	855 (150)

*Average of tests generally from several lots of material.

TABLE 5.4   Equations for Fracture Toughness Calculations

Case		Thicknesses appropriate to K_c	Plane strain conditions, K_{Ic}
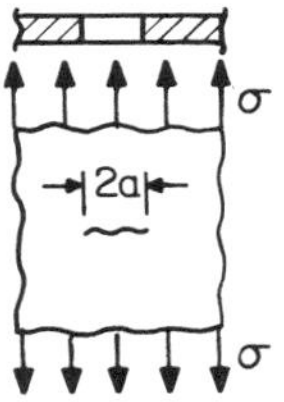	Through thickness crack in infinite plate	$\sigma = \dfrac{K_c}{\sqrt{\pi a}}$	$\sigma = \dfrac{K_{Ic}}{\sqrt{\pi a}}$
$\dfrac{2a}{w} < 0.2$	Through thickness crack in finite plate	$\sigma \simeq \dfrac{K_c}{\sqrt{\pi a}}$	$\sigma \simeq \dfrac{K_{Ic}}{\sqrt{\pi a}}$
$\dfrac{2}{3} a \leqslant 0.5t$	Surface crack	$\sigma \simeq \dfrac{1.14 K_c}{\sqrt{\pi a}}$	$\sigma \simeq \dfrac{1.14 K_{Ic}}{\sqrt{\pi a}}$
$\dfrac{a}{w} < 0.2$	Edge crack	$\sigma = \dfrac{0.8 K_c}{\sqrt{\pi a}}$	$\sigma = \dfrac{0.8 K_{Ic}}{\sqrt{\pi a}}$

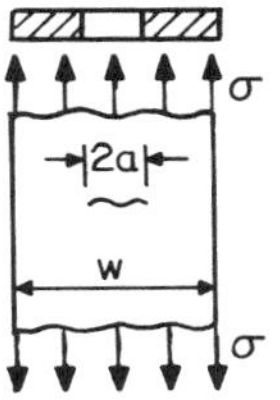

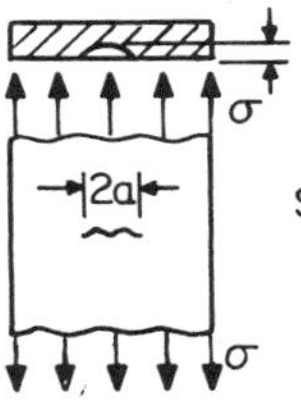

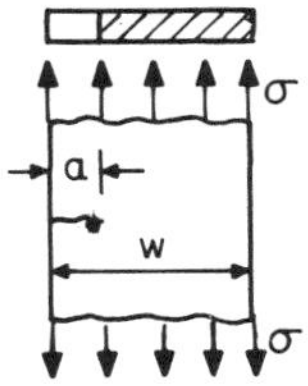

tical columns fail by buckling. The different types of column buckling are as follows:

Flexural buckling

Torsional buckling

Combined flexural-torsional buckling

Overall column buckling (as influenced by local element buckling)

The first three types are covered in this chapter. The assumption is that the shape of the cross section stays the same during overall buckling. The effect of element buckling on overall behavior (the last item in the list) is covered in Chap. 7.

5.2.1 Flexural buckling

The factors affecting flexural buckling of columns are listed below; they apply to all types of column buckling:

Alloy

Configuration/shape

Initial crookedness

End fixity

Welding/fabrication (reduced strength and residual stresses)

Other (load rate, temperature)

Alloy. The Euler buckling formula applies to all straight members. Modulus of elasticity is the only material property that must be known. For inelastic buckling the elastic modulus in the Euler equation may be replaced by the tangent modulus. Aluminum alloys can be placed into two groups, based on the shapes of their stress-strain curves. The two groups are those with temper designations starting with -T6, -T7, -T8, and -T9, and those with temper designations starting with -0, -H, -T1, -T2, -T3, and -T4. The inelastic column curve, based on the tangent modulus, is well approximated by a straight line.[6] Figure 5.8 shows the elastic curve and the two inelastic straight lines. The lower straight line is tangent to the elastic curve. The equations for column buckling are as follows:

$$\sigma = \frac{\pi^2 E}{\lambda^2} \tag{5.3}$$

where σ = column buckling strength
E = modulus of elasticity
λ = slenderness ratio (KL/r)
K = end-fixity coefficient
L = column length
r = radius of gyration

$$\sigma = B_c - D_c \lambda \tag{5.4}$$

where B_c, D_c = material parameters. The material constants for Eq. (5.4) are given in Table 5.5 along with the slenderness ratio, C_c that defines the intersection of the elastic and inelastic regions. Figure 5.9 shows the elastic and inelastic curves from Eqs. (5.3) and (5.4) and the

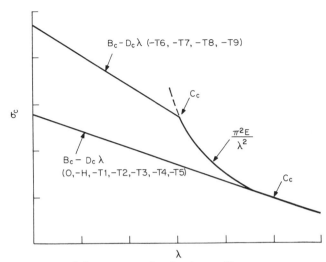

Figure 5.8 Column curves for aluminum alloys.

excellent agreement between calculated values and laboratory tests. The test columns were essentially straight.

Configuration/shape. The radius of gyration of the section in the direction of buckling is the important cross-sectional property in addition to area. The shape of the section affects strengths somewhat (usually by less than a few percent).[7,8] The effects are neglected in design. The results of tests of several shapes are included in Fig. 5.9 (each symbol denotes a shape). There are no significant differences in strength depending on shape.

Initial crookedness. Figure 5.10 shows load-deformation plots for a wide-flange column with no crookedness and for one having a crookedness of 1/1000 of the length, a reasonable value for a practical column. Columns that buckle at stresses equal to the proportional limit have the largest decrease in load due to crookedness. The proportional limit is given by the intersection of the elastic and inelastic column curves. For the studies referenced here[7] the proportional limit occurred at a slenderness of 67. Figure 5.11 shows how strength decreases with increase in crookedness. The effects are large enough that they need to be accounted for in design.

End fixity. The support details can have a large effect on the capacity of a column. The resistance to rotation and resistance to translation determine the appropriate K value. Some values are provided in Fig.

TABLE 5.5 Buckling Constants for Columns[6]

Temper	B_c		D_c		C_c
	ksi	MPa	ksi	MPa	
All alloys with tempers starting with -0, -H, -T1, -T2, -T3, -T4	$\sigma_y\left[1 + \left(\dfrac{\sigma_y}{1000}\right)^{1/2}\right]$	$\sigma_y\left[1 + \left(\dfrac{\sigma_y}{6895}\right)^{1/2}\right]$	$\dfrac{B_c}{20}\left(\dfrac{6B_c}{E}\right)^{1/2}$	$\dfrac{B_c}{20}\left(\dfrac{6B_c}{E}\right)^{1/2}$	$\dfrac{2B_c}{3D_c}$
All alloys with tempers starting with -T5, -T6, -T7, -T8, -T9	$\sigma_y\left[1 + \left(\dfrac{\sigma_y}{2250}\right)^{1/2}\right]$	$\sigma_y\left[1 + \left(\dfrac{\sigma_y}{15,510}\right)^{1/2}\right]$	$\dfrac{B_c}{10}\left(\dfrac{B_c}{E}\right)^{1/2}$	$\dfrac{B_c}{10}\left(\dfrac{6B_c}{E}\right)^{1/2}$	$\dfrac{0.41B_c}{D_c}$

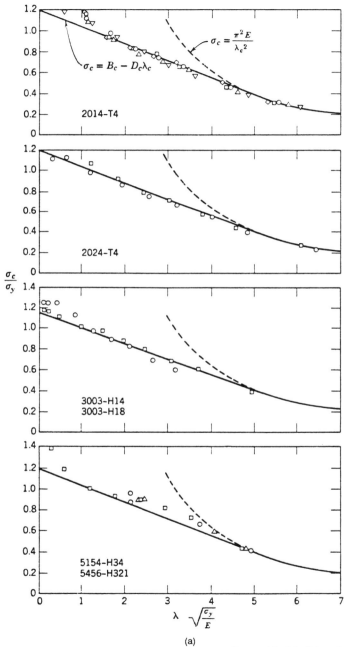

$$\sigma_c = \frac{\pi^2 E}{\lambda_c{}^2}$$

$$\sigma_c = B_c - D_c\lambda_c$$

2014–T4

2024–T4

$\dfrac{\sigma_c}{\sigma_y}$

3003–H14
3003–H18

5154–H34
5456–H321

$\lambda \sqrt{\dfrac{\sigma_y}{E}}$

(a)

Figure 5.9 Column strength of aluminum alloys with (a) -T4 and -H tempers and (b) -T6 tempers.[6]

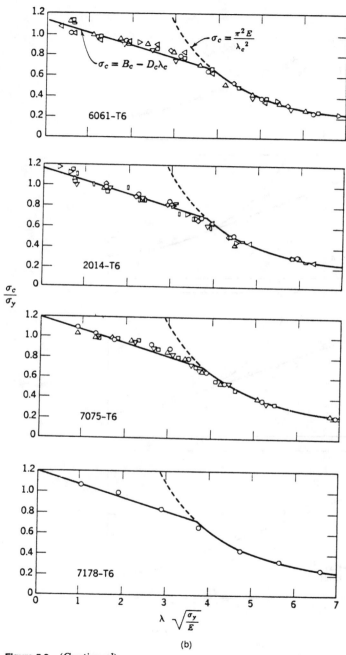

Figure 5.9   (*Continued*)

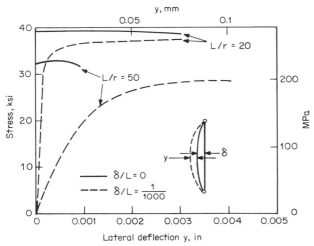

Figure 5.10   Behavior of 6061-T6 columns with crookedness.[7]

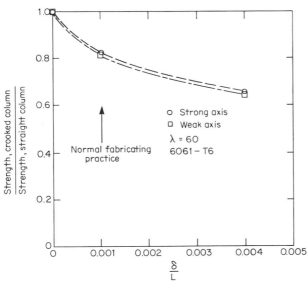

Figure 5.11   Effect of crookedness on column strength.[7]

5.12. The length of the column multiplied by this K value is the effective length of the column. There are some rational methods to calculate end fixity (see Ref. 9 for summary). Often the engineer must make an estimate because there are not very precise guidelines available. A fixed-end condition requires substantial, stiff framing and con-

Figure 5.12 Effective-length factors for centrally loaded columns.

	Flexural Buckling (Ref. 9)						Torsional Buckling		
Buckled shape of column is shown by dashed line	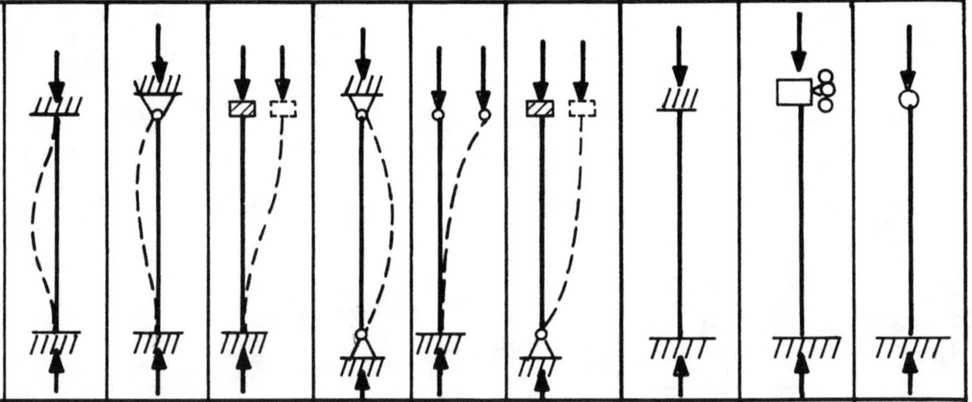								
Theoretical K value	0.5	0.7	1.0	1.0	2.0	2.0	0.5	1.0	2.0
Recommended K value when ideal conditions are approximated	0.65	0.80	1.2	1.0	2.10	2.0			
End condition code	 Rotation fixed. Translation fixed.	Rotation free. Translation fixed.	Rotation fixed. Translation free.	Rotation free. Translation free.			 End fixed against twist. Fixed against longitudinal warpage.	End fixed against twist. Free to warp.	End free to twist.

nections, and is difficult to achieve. In laboratory tests of aluminum specimens, fixed ends can be achieved by the use of carefully machined flat ends bearing on fixed steel plates. In contrast, the theoretical pinned end usually does not exist because some resistance is provided by almost all common attachments.

Figure 5.13 shows that a conservative assumption of end fixity compensates for the decrease in capacity due to crookedness. In this case a column with an effective length coefficient K of 0.9 and a crookedness of 0.001 L is as strong as or stronger than a straight column with a K of 1.0 (the horizontal dashed line). "Specifications for Aluminum Structures"[1] takes advantage of this trade-off by using a conservative value of K (equal to 1.0) and neglecting the crookedness effect.

Welding fabrication. Welding produces the reduced-strength material and the residual stresses discussed previously. For columns with longitudinal welds, the strength is given by an equation having the same form as that for a tension member:[10]

$$\sigma = \sigma_b - (A_r/A)(\sigma_b - \sigma_r) \tag{5.5}$$

where σ = column strength with longitudinal welds
$\quad\sigma_b$ = column strength assuming all base material
$\quad\sigma_r$ = column strength assuming all reduced-strength material

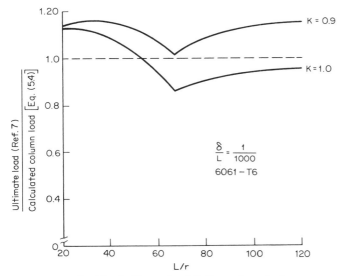

Figure 5.13 Combined effects of end fixity and crookedness.

In this case the column strengths for base and reduced strength materials are employed. Full-section properties are used for each calculation. Constants for inelastic buckling equations for reduced-strength materials are always based on the temper category (0, -H, T4, etc.), because the shape of the stress-strain curve for this material is no longer the same as the base material. Figure 5.14 shows that Eq. (5.5) predicts available test data well.[10] This equation is also reasonable compared to test results for larger columns with longitudinal welds.[8]

Transverse welds also affect column strength. Some of the data available[10] are shown in Fig. 5.15. Also shown are the column curves for base material and for as-welded material based on 10-in (254-mm) gage length yield strengths. Because most of the present test information is from small specimens and there is uncertainty as to how well the data apply to larger structures, other design approaches based on the length of the column affected by welding are not included. Thus, the design of columns with transverse welds can be based on the assumption that the entire column has the as-welded properties as provided in Table 5.2.

Other. If the column operates at elevated temperature, the material properties at that temperature need to be used in the buckling formulas. Loading at sustained times and temperatures can result in creep buckling. In addition, column strength is generally higher for rapidly applied loads compared to quasi-static loads. The strengthening is caused by dynamic effects (inertia resistances for example) and higher values for mechanical properties. These failure modes are not covered in this book.

5.2.2 Torsional buckling

Most columns are stiff enough torsionally so that they do not fail by twisting. However, some thin-walled, unsymmetrical sections, such as angles, tees, and cruciforms, may fail in this mode or a combined flexural-torsional mode. An equivalent slenderness ratio will be introduced for torsional, and for other buckling. The equivalent slenderness ratio is inserted into the elastic and inelastic column curves to calculate the torsional buckling stress. The equation is as follows:[11,12]

$$\lambda_\phi = \sqrt{\frac{I_x + I_y}{\dfrac{3}{8}\dfrac{J}{\pi^2} + \dfrac{C_w}{(K_\phi L)^2}}} \qquad (5.6)$$

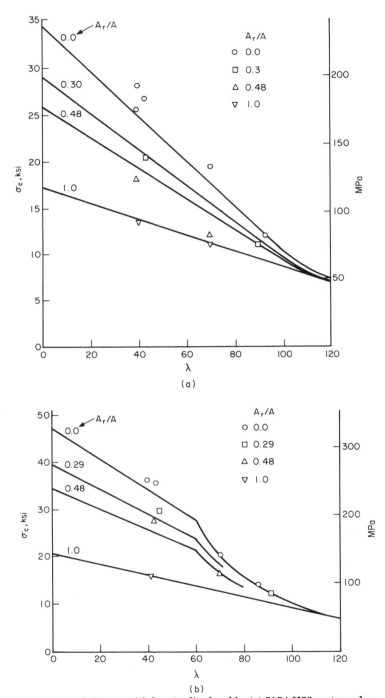

Figure 5.14 Columns with longitudinal welds. (*a*) 5154-H32 rectangular sections, (*b*) 6061-T6 rectangular sections, (*c*) 5456-H321 rectangular sections.[10]

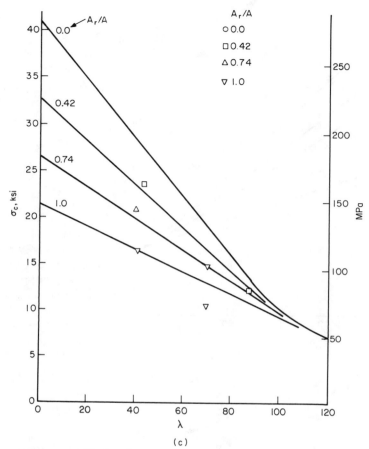

Figure 5.14   (*Continued*)

where λ_ϕ = equivalent slenderness ratio for torsional buckling

I_x, I_y = moments of inertia about the centroid in the x and y directions, respectively (principal axes)

J = torsion constant

C_w = warping constant

K_ϕ = effective length coefficient for torsional buckling

L = length of column

Figure 5.16 shows an unsymmetrical shape with definitions of terms that are needed for torsional and flexural-torsional buckling. The principal axes, the location of the shear center, and other dimensions are shown. Equations for warping and torsion properties are given in Table 5.6 and effective-length values are provided in Fig.

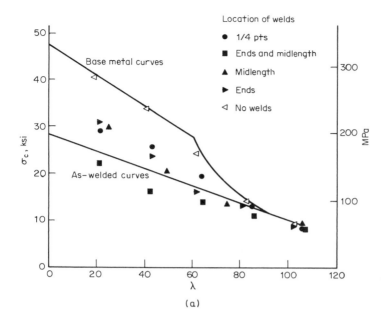

(a)

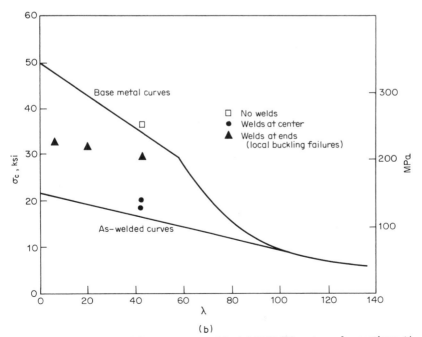

(b)

Figure 5.15 Columns with transverse welds. (a) 6061-T6 rectangular sections: ½ in (12.7 mm) × 2.0 in (50.8 mm), ¼ in (6.3 mm) × 2.0 in (50.8 mm). (b) 6061-T6 tubes: 3-in (76-mm) outside diameter, 0.25-in (6.4-mm) wall.[10]

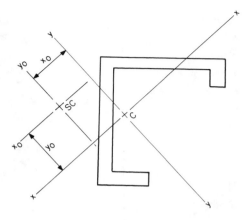

Figure 5.16   Definition of terms.

5.12. The effective lengths for torsion are not the same as those for flexural buckling because they are concerned with torsional, not bending, restraints.

5.2.3   Combined flexural-torsional buckling

Thin, unsymmetrical shapes such as that shown on Fig. 5.16 will probably fail by torsional-flexural buckling. The solution has already been developed.[11] The equation can be put in the following form:

$$\left[1 - \left(\frac{\lambda_c}{\lambda_y}\right)^2\right]\left[1 - \left(\frac{\lambda_c}{\lambda_x}\right)^2\right]\left[1 - \left(\frac{\lambda_c}{\lambda_\phi}\right)^2\right]$$

$$- \left(\frac{y_o}{r_o}\right)^2\left[1 - \left(\frac{\lambda_c}{\lambda_x}\right)^2\right] - \left(\frac{x_o}{r_o}\right)^2\left[1 - \left(\frac{\lambda_c}{\lambda_y}\right)^2\right] = 0 \quad (5.7)$$

where λ_c = equivalent slenderness ratio for torsional-flexural buckling

λ_x, λ_y = effective slenderness ratios for flexural buckling in the x and y directions, respectively

λ_ϕ = equivalent slenderness ratio for torsional buckling

x_o, y_o = distances between centroid and shear center, parallel to principal axes

$r_o = \sqrt{(I_{x_o} + I_{y_o})/A}$

I_{x_o}, I_{y_o} = moments of inertia about axes through shear center

A = area of section

Equation (5.7) can be solved by trial. For sections with double symmetry the centroid and shear center coincide. The right two terms are zero and the solution is the largest slenderness ratio for flexural buck-

TABLE 5.6 Warping Constants for Thin-Walled Shapes*†

Case	Location of shear center e	Warping constant C_w	References
		$\dfrac{d^2 I_y}{4}$	8, 12, 13
		$\dfrac{d^2 I_y}{4} + c^2 b^2 t\left(\dfrac{d}{2} + \dfrac{c}{3}\right)$	8
	$\dfrac{y_1 I_1 - y_2 I_2}{I_y}$	$\dfrac{d^2 I_1 I_2}{I_y}$	8, 12

NOTE: See page 87 for footnotes.

TABLE 5.6 Warping Constants for Thin-Walled Shapes*† (Continued)

Case	Location of shear center e	Warping constant C_w	References
	$\dfrac{xd^2}{4r_x^2}$	$\dfrac{d^2I_y}{4}\left[1 - \dfrac{x(a-x)}{r_y^2}\right]$	12
	$\dfrac{d^2b^2t}{I_x}\left(\dfrac{1}{4} + \dfrac{c}{2b} - \dfrac{2c^3}{3d^2b}\right)$	$\dfrac{b^2t}{6}(4c^3 + 6dc^2 + 3d^2c + d^2b) - e^2I_x$	8
	$\dfrac{d^2b^2t}{I_x}\left(\dfrac{1}{4} + \dfrac{c}{2b} - \dfrac{2c^3}{3d^2b}\right)$	$\dfrac{b^2t}{6}(4c^3 - 6dc^2 + 3d^2c + d^2b) - e^2I_x$	8

$$\frac{d^2 I_y}{4}\left(1 - \frac{3A_F}{2A}\right)$$

12

$$\frac{b^2 t}{12(2b + d + 2c)}\left[d^2(b^2 + 2bd + 4bc + 6dc) + 4c^2(3db + 3d^2 + 4bc + 2dc + c^2)\right]$$

8

See sketch

0

12

The sketches of the second, fifth, sixth, and eighth cases are from Fig. 7.81 of *Aluminium Alloy Structures* by F. M. Mazzolani (Pitman Publishing Company, London, 1985).

*I_y = moment of inertia of section about y axis, I_1 = moment of inertia of flange No. 1 about y axis, I_2 = moment of inertia of flange No. 2 about y axis, r_y = radius of gyration of section about y axis, r_x = radius of gyration of section about x axis, A_F = area of one flange, A = total area of section.

†Torsion constant J for thin-walled sections shown here is given by $J = \Sigma \frac{1}{3}at^3$. This assumes that the section consists of rectangular elements of length a and thickness t, and that the contributions of all segments are added together to estimate J.

ling in the x and y directions or for torsional buckling. One of the right-hand terms will go to zero for singularly symmetrical sections. The slenderness ratio for flexural-torsional buckling will be larger than those for pure flexural or torsional buckling. Figure 5.17 shows data from Ref. 14 and calculations using Eq. (5.7). The ends of the specimens were machined flat and restrained against rotation in the tests. Fixed-end conditions were assumed except for the warpage term, in which restrained and unrestrained values were used. The results are in between these limits but nearer to the fixed condition. Equation (5.7) is reasonable based on these data.

5.2.4 Summary of column design

The above discussion covers most of the common types of column buckling failure, with flexural buckling being the most common. One set of column curves can be used. No factors of safety are included in any of the formulas. Thus, uncertainties such as crookedness and end conditions must be accommodated by an appropriate conservativeness in the design.

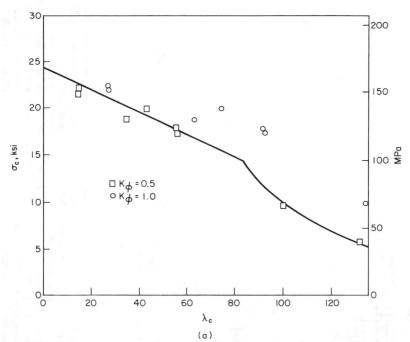

(a)

Figure 5.17   Flexural-torsional buckling. (a) 6063-T6 lipped channels, (b) 6061-T6 hat sections, (c) 5052-H34 hat sections, (d) 6061-T6 TEES.[14]

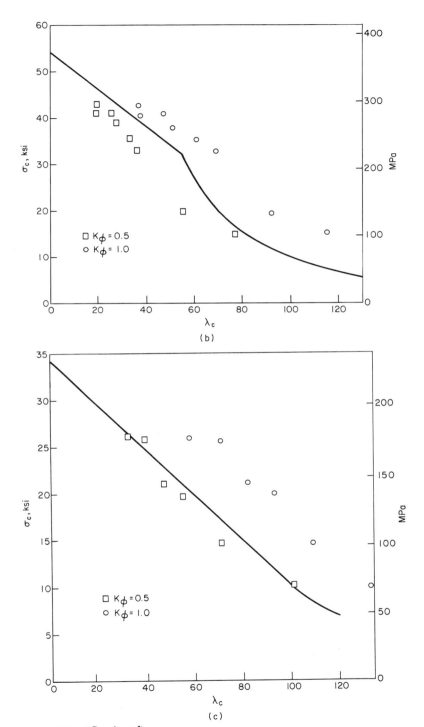

Figure 5.17 *(Continued)*

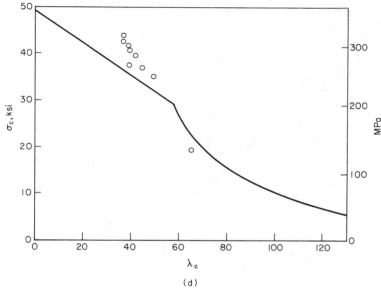

Figure 5.17 (*Continued*)

References

1. "Specifications for Aluminum Structures," *Aluminum Construction Manual*, Sec. 1, The Aluminum Association, Washington, D.C., 1986.
2. Hill, H. N., Clark, J. W., and Brungraber, R. J., "Design of Welded Aluminum Structures," paper presented at the Annual Meeting of the American Society of Civil Engineers, Washington, D.C., October 19–23, 1959.
3. Kaufman, J. G., "Fracture Toughness of Aluminum Alloy Plate from Tension Tests of Large Center-Slotted Panels," reprint from *Journal of Materials,* Vol. 2, No. 4, American Society for Testing and Materials, 1967.
4. Kaufman, J. G., and Holt, Marshall, "Fracture Characteristics of Aluminum Alloys," Alcoa Research Laboratories Technical Paper No. 18, Alcoa Center, Pennsylvania, 1965.
5. Rolfe, Stanley T., and Barsom, John M., *Fracture and Fatigue Control in Structures—Applications of Fracture Mechanics,* Prentice-Hall, Englewood Cliffs, New Jersey, 1977.
6. Clark, J. W., and Rolf, Richard L., "Buckling of Aluminum Columns, Plates and Beams," *Proceedings of the American Society of Civil Engineers,* Journal of the Structural Division, June 1966.
7. Batterman, Richard H., and Johnston, Bruce G., "Behavior and Maximum Strength of Metal Columns," *Proceedings of the American Society of Civil Engineers,* Journal of the Structural Division, April 1967.
8. Mazzolani, F. M., *Aluminum Alloy Structures,* Pitman, Marshfield, Massachusetts, 1985.
9. Galambos, Theodore V. (ed.), *Guide to Stability Design Criteria for Metal Structures,* 4th ed., Wiley, New York, 1988.
10. Brungraber, R. J., and Clark, J. W., "Strength of Welded Aluminum Columns," *Proceedings of the American Society of Civil Engineers,* Journal of the Structural Division, August 1960.
11. Timoshenko, S. P., and Gere, J. M., *Theory of Elastic Stability,* 2d ed., McGraw-Hill, New York, 1961.

12. *Alcoa Structural Handbook*, Aluminum Company of America, Pittsburgh, 1960.
13. Marsh, Cedric, *Strength of Aluminum*, 5th ed., The Aluminum Company of Canada, Limited, Montreal, Canada, 1983.
14. Abramson, Andrew B., "Inelastic Torsional-Flexural Buckling of Aluminum Sections," Report No. 365, Department of Structural Engineering, School of Civil and Environmental Engineering, Cornell University, Ithaca, New York, October 1977.

6

Bending Members and Beam Columns

Aluminum components or members subjected to bending, torsion, and combined bending and axial loads are considered in this chapter. The discussion is confined to sections in which the elements do not distort during loading. Buckling of elements and its effect on component behavior is covered in Chap. 7. Failure modes for beams laterally supported along the length are yielding or tensile fracture, similar to that described for tension members. The difference between tensile and bending members is that bending members can accommodate loads much higher than those calculated by elastic analysis. A "shape factor" is needed for a better estimate of load-carrying capacity. A similar effect results from loads causing torsion.

For beams that do not have continuous lateral support, the component between points of lateral support may fail by lateral buckling, an important failure mode for bending members. Finally, the failure of bending members is greatly affected by the presence of axial loads. The combined effect is handled by an interaction equation.

6.1 Bending Components Failing by Yielding and Fracture

6.1.1 Members without welds

The extreme fiber stress is given by the flexure formula:

$$\sigma = \frac{M}{S} \tag{6.1}$$

where σ = stress (usually extreme fiber stress)
S = section modulus for the point of interest
M = applied bending moment

The moment and section modulus are calculated by any of the established structural mechanics techniques. However, Eq. (6.1) would give overly conservative results if the calculated stress were limited to the yield or the tensile strength of the alloy. The normal design procedure is to introduce a shape factor that is used to "increase" the strength of the alloy so that a more accurate, less conservative, capacity of the beam is calculated. Figure 6.1 illustrates the basis for calculation of the shape factor. The strain distribution is assumed to vary linearly through the depth of any section for elastic and inelastic strains. Also shown are stress distributions for an aluminum section with initial yielding at the extreme fiber, general yielding, and ultimate strength. Obviously, much higher moments than those based on initial yielding are possible depending on the shape of the stress-strain curve and section; the moment capacity depends on the areas of the stress-distribution curves.

Figure 6.2 shows an axial stress-strain curve for 6061-T6, and the calculated curve for bending of a rectangular shape based on the inelastic distribution of stress in the section. Yielding in the beam is assumed to occur at the 0.2 percent offset in strain, the same definition of yielding as that employed for axially loaded parts. In this case the calculated elastic stress is 1.32 times the uniaxial yield and the ultimate strength is 1.46 times the uniaxial tensile strength. In making the calculations for bending, the strain distribution is assumed to be linear as shown in Fig. 6.1 and the stress-strain relationship is considered to be the same as that for the axial stress test. The conservative assumption is made that the compressive stress-strain curve is the same as the tensile curve. The bending moment M corresponding to a given extreme fiber strain is calculated by numerical integration, assuming a rectangular-shaped cross section, and the "apparent bending stress," M/S, corresponding to the given strain is calculated. Calculated values using this method are in reasonable agreement with test data as illustrated in Fig. 6.3. The shape factors calculated for these two cases are less than the shape factor of 1.5 that applies to a rigid-plastic material (assumed for mild steel) because of the shape of

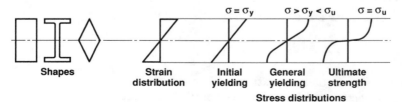

Figure 6.1 Inelastic bending of aluminum shapes.

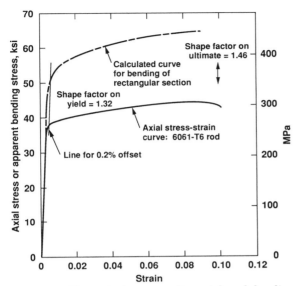

Figure 6.2 Stress-strain curves for axial and bending members.

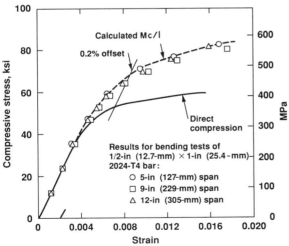

Figure 6.3 Bending of rectangular beams.[1]

the stress-strain curve. In addition, the shape factor for ultimate strength is higher than and nearer to the rigid-plastic case compared to the value for yielding. Finally, although the stress-strain curves are somewhat different for the various alloys, the shape factors are close enough so that one value may be used for design.

For design purposes, it is much easier to calculate the rigid-plastic values than values based on specific aluminum stress-strain curves. Table 6.1 and Fig. 6.4 show that values for aluminum may be estimated from the rigid-plastic cases using the following relationships:

For yielding: $\qquad$ $Z = 0.4 + 0.6Z_p$ $\qquad$ (6.2)

For ultimate strength: $\quad$ $Z = 0.2 + 0.8Z_p$ $\qquad$ (6.3)

where Z = shape factor for aluminum alloys
$\quad$ Z_p = shape factor for rigid-plastic case

In the calculation of beam strength the yield or tensile strength of the alloy is multiplied by the shape factor from Eqs. (6.2) and (6.3), to obtain the stress to compare with the elastic stress calculated from Eq. (6.1).

TABLE 6.1 Shape Factors for Bending*

Shape	Theoretical, rigid-plastic stress-strain curve	Theoretical calculations, 6061-T6 rod	
		Yield	Ultimate
I-beam, $t \ll b$	1.0	1.0	1.0
I-beam, $b/10$, $b/20$	1.15	1.13	1.14
rectangle	1.5	1.32	1.46
diamond	2.0	1.62	1.72

*Moment = shape factor times yield or ultimate strength times section modulus.

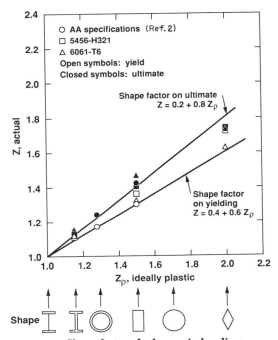

Figure 6.4 Shape factors for beams in bending.

6.1.2 Members with welds

The average strength of the material in beams with longitudinal welds is determined in the same manner as the weighted-average strength for tensile members. The area of the beam flange to be used in the calculations consists of the portion of the beam further than $2c/3$ from the neutral axis, where c is the distance from the neutral axis and the extreme fiber.

The same shape factors given above may be applied to beams with longitudinal welds, but may be unconservative for beams with transverse welds at the point of maximum moment, particularly for ultimate load. Test data for beams of several alloys are provided in Fig. 6.5. The shape factors for ultimate strength for beams with no welds or longitudinal welds were within 8 percent of calculated values, while those for beams with transverse welds were as much as 20 percent lower than calculated values. It is conservative to use a shape factor of 1.0 for these cases.

6.1.3 Moment redistribution in frames

There have been only limited studies to investigate the ability of continuous aluminum frames to redistribute moments in the manner as-

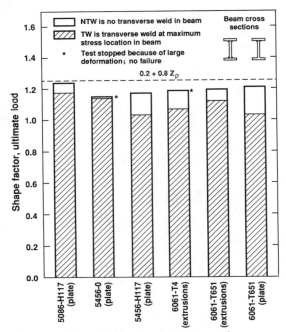

Figure 6.5 Shape factors for welded beams.[3]

sumed in "plastic design."[4,5] Thus, there is not sufficient work to establish design techniques. This discussion is limited to some of the information that is available that can be applied if the designer chooses to conduct structural simulations of specific cases. One of the requirements is that the elements of the section are sufficiently compact that large rotations may occur without buckling of the elements. Figure 6.6 presents data for buckling failures of flanges of beams. Also shown is the calculated shape factor for ultimate strength. None of the sections tested was compact enough to develop the calculated shape factor on ultimate strength, or in other words, the expected moment capacity of the section. The vertical lines are limits proposed elsewhere for compact sections based on considerations of rotational capacity.[6] The proposed limits for flanges appear to be satisfactory. There are not enough data available at this time to confirm limits for other elements under compression or for limits of the slenderness of beams to prevent lateral buckling.

Aluminum beams can achieve large plastic distortions and rotations as illustrated in Fig. 6.7. Shown is the test of a 6061-T4 member, which was stopped because there was no more room for deflection in the fixtures.[3] The deformation capacity of aluminum beams with com-

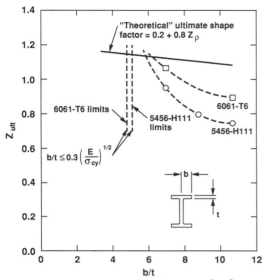

Figure 6.6 Proportions of sections to develop expected ultimate moment.

Figure 6.7 Aluminum beam with large plastic deformations.[3]

pact sections is highly dependent on alloy and temper and on the type of weld or joint at the point of maximum moment.[3] Although there are some references available concerning the redistribution of moments in aluminum continuous beams or frames,[4,5] there are no simplified plastic design techniques similar to those proposed for mild steel structures. A complication for aluminum design is that transverse welds in shapes of most alloys would limit the rotations that could occur because the strength of the weld is less than the base-metal strength, or in some cases less than the base-metal yield strength. There is little doubt that aluminum continuous frames can achieve strengths above that corresponding to initial yielding.[4] However, a method for an accurate assessment of strength needs to take into account the characteristics of the alloy as well as influence of welds or other joints.

6.2 Components Failing in Torsion

The capacity of a member loaded in torsion is dependent on the cross section and on the shear yield and ultimate strengths of the alloy. The elastic stresses for several shapes are provided in Table 6.2.[7] Similar to the behavior discussed for bending members, the formulas no longer apply when the stresses exceed the yield strength of the material. The equations given for ultimate torque are based on the assumption that the shear ultimate strength of the material is reached on each part of the cross section. Based on the work on bending members (see Fig. 6.4, for example), the assumption is somewhat unconservative but probably less than 10 percent in error for most shapes. Table 6.2 covers shapes that are efficient in transmitting torsional loads.

Although inefficient in transmitting torsional loads, open sections such as channels and wide-flange sections may also be loaded in torsion. The stresses in these parts are much more complicated to calculate because warping of the section needs to be taken into account. Established strength-of-materials solutions or computer simulations are options for analysis.

6.3 Beams Failing by Lateral Buckling

The factors that affect the lateral buckling of a beam are similar to those affecting flexural buckling of a column:

- Configuration/shape
- Type and location of load
- End fixity
- Alloy (in the inelastic range of the material)

TABLE 6.2 Torsion Formulas[7]

Maximum $\tau = \dfrac{2T}{\pi R^3}$ at boundary

$$T_U = \dfrac{2 \pi R^3 \tau_U}{3}$$

$$J = \dfrac{\pi R^4}{2}$$

Maximum $\tau = \dfrac{3T}{bt^2}\left(1 + 0.6\,\dfrac{t}{b}\right)$ at midpoints of long sides

(approximately $\dfrac{3T}{bt^2}$ for narrow rectangles)

$$T_U = \dfrac{bt^2 \tau_U}{2}\left(1 - \dfrac{t}{3b}\right)$$

$$J = \dfrac{bt^3}{3}\left[1 - 0.63\,\dfrac{t}{b} + 0.052\left(\dfrac{t}{b}\right)^2\right]$$

(approximately $\dfrac{bt^3}{3}$ for narrow rectangles)

Maximum $\tau = \dfrac{2R_1 T}{\pi\,(R_1^4 - R_2^4)}$ at boundary

$$T_U = \dfrac{2\pi(R_1^3 - R_2^3)\tau_U}{3}$$

$$J = \dfrac{\pi}{2}\,(R_1^4 - R_2^4)$$

Thin-walled sections should be checked for buckling

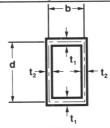

For sides with thickness t_1,

average $\tau = \dfrac{T}{2t_1 bd}$

For sides with thickness t_2,

average $\tau = \dfrac{T}{2t_2 bd}$

$$T_U = 2bd\, t_{min}\, \tau_U$$

($t_{min} = t_1$ or t_2, whichever is smaller)

$$J = \dfrac{2b^2 d^2}{\dfrac{b}{t_1} + \dfrac{d}{t_2}}$$

Thin-walled sections should be checked for buckling

Maximum $\tau = \dfrac{4.8T}{a^3}$ at midpoints of sides

$$T_U = \dfrac{a^3 \tau_U}{3}$$

$$J = 0.141 a^4$$

τ = shear stress (formulas for τ apply for stress not exceeding the shear yield strength), T = torque, T_u = approximate ultimate torque, τ_u = ultimate shear strength, J = torsion constant.

- Initial crookedness
- Welding/fabrication

Configuration/shape. This factor is considered first because it will affect the manner in which buckling in the inelastic range is handled. An *equivalent slenderness ratio* is introduced, which will be used in elastic and inelastic buckling curves in the same way that the slenderness ratio of a column is used. A critical moment is calculated for lateral buckling of a beam based on an elastic solution. The critical stress in the beam is calculated as follows:

$$\sigma_c = \frac{M_c}{S_x} \tag{6.4}$$

where σ_c = critical stress
M_c = critical moment
S_x = section modulus corresponding to the compression flange about the axis of bending

This critical stress is equated to the elastic buckling stress,

$$\sigma_c = \frac{\pi^2 E}{\lambda_e^2} \tag{6.5}$$

where E = modulus of elasticity
λ_e = equivalent slenderness ratio

Combining Eqs. (6.4) and (6.5) and solving for an equivalent slenderness ratio provides the following:

$$\lambda_e = \pi \sqrt{\frac{ES_x}{M_c}} \tag{6.6}$$

The equivalent slenderness ratio for a beam symmetrical about the axis of bending and loaded through the shear center is as follows:[1]

$$\lambda_e = \frac{k_b(KL)\sqrt{S_x}}{\left\{I_y\left[C_w + \dfrac{G}{\pi^2 E} J(KL)^2\right]\right\}^{1/4}} \tag{6.7}$$

where k_b = coefficient depending on type of load and configuration (Table 6.3, Figs. 6.8 and 6.9)
K = coefficient for end fixity (rotation about vertical axis)
L = span length
S_x = section modulus of compression flange
I_y = moment of inertia (lateral direction)
G = $E/2$ $(1 + \nu)$ = modulus of rigidity
C_w = warping constant (Table 5.6)
J = torsion constant (Table 6.2)
E = modulus of elasticity
ν = Poisson's ratio ($\frac{1}{3}$)

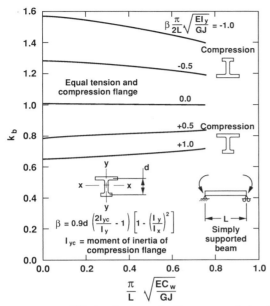

Figure 6.8 Effect of unequal flanges on lateral buckling.[8]

The equations given thus far have applied to sections with equal sizes of the tensile and compressive flanges. Figure 6.8 shows the variation in the buckling coefficient as the relative proportions of the flanges vary. A larger compression flange relative to the tensile flange is beneficial for higher buckling values.

Type and location of loads. Constants for various loading cases and end conditions are provided in Table 6.3. The constants apply to cases in which the load is applied at the shear center. Lateral buckling is affected by the location of the load on the beam as illustrated in Fig. 6.9. The strongest arrangement is for the case in which the load is hung from the bottom (tensile) flange.

End fixity. Table 6.3 also has values for the end fixity at the point of support corresponding to free or fixed against rotation about the vertical axis. The ends are considered to be free from warping restraints except for cantilevers.

One of the common uncertainties in applying Eq. (6.7) is the definition of the unsupported length. The length L is the distance between points of the beam that are supported against lateral movement. Points of inflection of the beam do not qualify as points of lateral sup-

TABLE 6.3 Coefficients for Lateral Buckling of Beams[9]

Loading	Bending moment diagram	Condition of restraint against rotation about vertical axis at ends	K	k_b
(L) with M	M	Simple support	1.0	1.0
		Fixed	0.5	1.0
M … $M/2$	M … $M/2$	Simple support	1.0	0.87
		Fixed	0.5	0.87
M	M	Simple support	1.0	0.75
		Fixed	0.5	0.75
M … $M/2$	M … $M/2$	Simple support	1.0	0.66
		Fixed	0.5	0.66
M … M	M … M	Simple support	1.0	0.62
		Fixed	0.5	0.62
W	$WL/8$	Simple support	1.0	0.94
		Fixed	0.5	1.02
W	$2\,M_{cr}$; $WL/24 = M_{cr}$	Simple support	1.0	0.88
		Fixed	0.5	1.08
P	$PL/4$	Simple support	1.0	0.86
		Fixed	0.5	0.97
P	$PL/8$; $PL/8$	Simple support	1.0	0.77
		Fixed	0.5	0.98
$P/2$ $P/2$; $L/4$	$PL/8$	Simple support	1.0	0.98
P	PL	Warping restrained at supported end	1.0	0.88
W	$WL/2$	Warping restrained at supported end	1.0	0.70

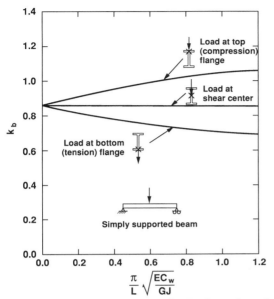

Figure 6.9 Effect of position on load on lateral buckling.[8]

port. A conservative estimate of the buckling strength for other moment distributions may be obtained by selecting a known distribution that encompasses the moment distribution for which a solution is not available. Both the maximum moment near the center of the beam and the maximum moment at the support need to be considered.

Alloy. The equivalent slenderness ratio for lateral buckling may be used in the elastic and inelastic column-buckling equations. This procedure is reasonable for wide-flange and similar shapes because their shape factor is generally close to 1.0, but it would be overly conservative for sections with higher shape factors when considering buckling in the inelastic region. Figure 6.10 shows a comparison of calculated strengths with test data for I-sections. The use of column curves for the inelastic range is conservative. The buckling constants are reproduced in Table 6.4. Also given in Table 6.4 are inelastic buckling constants for rectangular sections. These constants were developed from the beam-bending stress-strain relationships.[1] Figure 6.11 shows that the inelastic curve as provided by these constants is conservative compared to test data. Figure 6.11 also shows that the column curve would be grossly conservative for these beams.

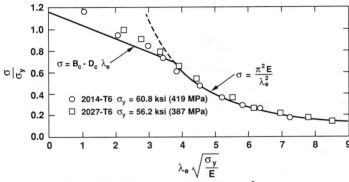

Figure 6.10   Lateral buckling strength of I-beams.[1]

Initial crookedness.   The effect of initial crookedness has not been explored for aluminum beams to the extent that it has been for aluminum columns. However, the effects should be similar for buckling of columns and beams; thus, the discussion in the preceding chapter should apply in a qualitative manner.

Welding/fabrication.   There has been little work on the lateral buckling strength of aluminum beams with various types of welds. Presumably the same method used for columns can be used for beams to develop the inelastic buckling curves. The area of the compression flange as defined above for the tensile flange is used in the calculations.

6.4   Components Failing by Combined Bending and Compression

There are two major conditions that need to be considered when the component has both axial and bending loads; the member can fail by bending in the plane of the applied moments or it can fail by lateral buckling. Fortunately, both modes of failure can be handled with the same interaction formula:[10,11]

$$\frac{P}{P_u} + \frac{C_m M}{M_u\left(1 - \dfrac{P}{P_e}\right)} \le 1 \tag{6.8}$$

where P, M = applied axial load and bending moments
  P_u, M_u = ultimate column and beam bending strength
  P_e = Euler buckling strength = $\pi^2 EI/L^2$
  C_m = coefficient for cases of nonuniform moment

TABLE 6.4 Buckling Constants for Beams[1]

Type of beam	Temper	B_c (ksi)	B_c (MPa)	D_c (ksi)	D_c (MPa)	C_c
Wide-flange I-sections	All alloys with tempers starting with -0, -H, -T1, -T2, -T3, -T4	$\sigma_y\left[1 + \left(\dfrac{\sigma_y}{1000}\right)^{1/2}\right]$	$\sigma_y\left[1 + \left(\dfrac{\sigma_y}{6895}\right)^{1/2}\right]$	$\dfrac{B_c}{20}\left(\dfrac{6B_c}{E}\right)^{1/2}$	$\dfrac{B_c}{20}\left(\dfrac{6B_c}{E}\right)^{1/2}$	$\dfrac{2}{3}\dfrac{B_c}{D_c}$
	All alloys with tempers starting with -T5, -T6, -T7, -T8, -T9	$\sigma_y\left[1 + \left(\dfrac{\sigma_y}{2250}\right)^{1/2}\right]$	$\sigma_y\left[1 + \left(\dfrac{\sigma_y}{15510}\right)^{1/2}\right]$	$\dfrac{B_c}{10}\left(\dfrac{B_c}{E}\right)^{1/2}$	$\dfrac{B_c}{10}\left(\dfrac{6B_c}{E}\right)^{1/2}$	$\dfrac{0.41B_c}{D_c}$

Type of beam	Temper	B_b (ksi)	B_b (MPa)	D_b (ksi)	D_b (MPa)	C_b
Rectangular section	All alloys and tempers	$1.3\sigma_y\left[1 + \dfrac{(\sigma_y)^{1/3}}{7}\right]$	$1.3\sigma_y\left[1 + \dfrac{(\sigma_y)^{1/3}}{13.3}\right]$	$\dfrac{B_p}{20}\left(\dfrac{6B_p}{E}\right)^{1/2}$	$\dfrac{B_p}{20}\left(\dfrac{6B_p}{E}\right)^{1/2}$	$\dfrac{2}{3}\dfrac{B_p}{D_p}$

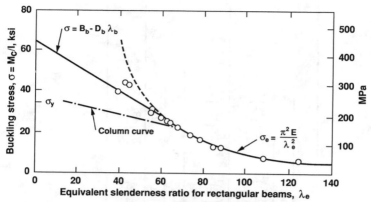

Figure 6.11   Buckling of 2017-T4 rectangular beams.[1]

The coefficient C_m allows for unequal moments at the ends of the components. The equation follows. Figure 6.12 also provides an interpretation of this coefficient.

$$C_m = 0.6 + 0.4 \frac{M_1}{M_2} \geq 0.4 \tag{6.9}$$

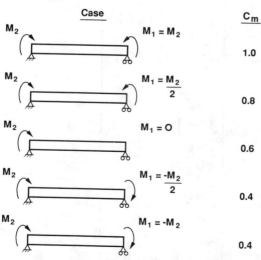

Figure 6.12   Interpretation of C_m in interaction formula.

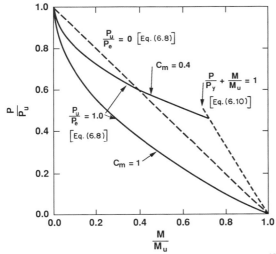

Figure 6.13 Interaction equations for beam columns.[10]

where M_1, M_2 = moments at the two ends of the member (M_1 is the smallest numerically of the moments).

Finally, another interaction equation is needed for those cases in which high moments at the end can cause excessive yielding. The equation is as follows:

$$\frac{P}{P_y} + \frac{M}{M_u} \leq 1.0 \qquad (6.10)$$

where P_y = area of column times the compressive yield strength.

Figure 6.13 shows extreme values for the interaction Eqs. (6.8) and (6.10). Note that the line for Eq. (6.10) is arbitrarily shown because it depends on the specific section and alloy.

The development of the above interaction equations is semi-empirical and much more background is available elsewhere.[8,11] The equations also apply to all metallic components. Equation (6.8) has an amplification term in the bending part $[1 - (P/P_e)]$ to take into account the effect of the additional moments caused by axial load and the deflection of the components due to the primary bending moments.

Test data for aluminum components are available for uniform bending only.[11] Figure 6.14 shows the data and the reasonable agreement between tests and calculation procedures.

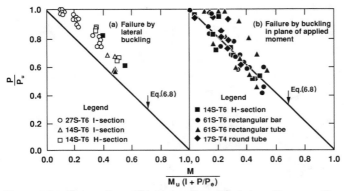

Figure 6.14 Comparison of test results with interaction equation (6.8). (Solid symbols indicate inelastic failure.)[11]

References

1. Clark, John W., and Rolf, Richard L., "Buckling of Aluminum Columns, Plates and Beams," *Proceedings of the American Society of Civil Engineers,* Journal of the Structural Division, June 1966.
2. "Specifications for Aluminum Structures," *Aluminum Construction Manual,* The Aluminum Association, Washington, D.C., 1986.
3. Sharp, M. L., "Static and Dynamic Behavior of Welded Aluminum Beams," *Welding Journal Research Supplement,* February 1973.
4. Potts, Richard G., and Brungraber, Robert J., "Inelastic Behavior of Structural Frameworks," *Proceedings of the American Society of Civil Engineers,* Journal of the Structural Division, June 1967.
5. Mazzolani, F. M., *Aluminum Alloy Structures,* Pitman, Marshfield, Massachusetts, 1985.
6. Gaylord, E. H., and Gaylord, C. N., "Design of Aluminum Structural Members," *Structural Engineering Handbook,* Sec. 10, McGraw-Hill, New York, 1988.
7. *Alcoa Structural Handbook,* Aluminum Company of America, Pittsburgh, 1960.
8. Galambos, Theodore V. (ed.), *Guide to Stability Design Criteria for Metal Structures,* 4th ed., Wiley, New York, 1988.
9. Clark, J. W., and Hill, H. W., "Lateral Buckling of Beams," *Journal of the Structural Division, Proceedings of the American Society of Civil Engineers,* July 1960.
10. Chapuis, Jacques, and Galambos, Theodore V., "Reliability of Aluminum Beam-Columns," *Proceedings of the American Society of Civil Engineers,* Journal of the Structural Division, April 1982.
11. Hill, Harry N., Hartmann, Earnest C., and Clark, J. M., "Design of Aluminum Alloy Beam-Columns," *Transactions, ASCE,* Vol. 121, 1956, p. 1.

Flat-Plate Elements

Plates supported along at least one of the unloaded edges and subjected to various types of in-plane loadings are considered in this chapter. The problems of design generally involve buckling under compression or shear, not tension. The design for tensile loads normally is associated with fastening methods, the subject covered in Chap. 10.

The initial portion of this chapter covers the classical buckling of plates that are long with respect to their width, representative of an element of a wide-flange or box member. Various support conditions and types of loading are handled by means of an equivalent slenderness ratio, the same technique employed for the lateral buckling of beams. Straight-line segments are provided for buckling in the inelastic range of the material.

The latter portions of the chapter are concerned with the postbuckling behavior of plates; the strength of plates is often much higher than the buckling value. Finally, methods for calculating failure when various types of load are on a single plate are provided.

The behavior of a plate for one of the cases considered is illustrated in Fig. 7.1. The long plate is loaded in uniaxial compression. For a perfectly flat plate, no lateral deflection occurs until the buckling value. After buckling, the plate deflects laterally and is able to carry higher loads because the stresses redistribute with the two longitudinal edges carrying much higher stresses than the interior portion of the plate. The deflected shape is a series of in-and-out buckles with the length of the half-wave of buckling about equal to the width of the plate. The maximum stress and load occur when the edge stresses are about equal to the yield strength of the alloy. As noted, the practical plate with initial crookedness exhibits no well-defined point of buck-

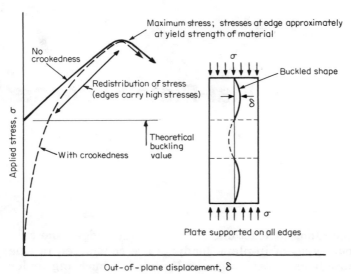

Figure 7.1 Behavior of plate under compressive stresses.

ling, but it has an ultimate strength essentially the same as that for the perfectly flat plate.

7.1 Buckling of Plates

The factors that affect the buckling of aluminum plates are similar to those defined for columns and beams:

Configuration/shape

Type of load

Fixity of the edges

Alloy and temper

Initial crookedness

Welding/fabrication

The first three can be handled by the use of an equivalent slenderness ratio, which, in a manner similar to the procedure used for the lateral buckling of beams, is inserted in equations for elastic and inelastic buckling of plates to obtain buckling strength. The equations for inelastic buckling are based on an effective modulus of elasticity that is higher than the tangent modulus used for columns and for some beams, because of the twisting and bending distortions that occur in plate buckling, compared to the predominantly bending distortions that occur in column buckling.[1]

7.1.1 Equivalent slenderness ratio

The elastic buckling stress for plates is usually expressed as follows:

$$\sigma = \frac{k\pi^2 E}{1 - \nu^2} \left(\frac{t}{b}\right)^2 \tag{7.1}$$

where σ = stress causing buckling of the plate
$\quad k$ = coefficient that depends on type of load applied and configuration of plate
$\quad E$ = modulus of elasticity
$\quad \nu$ = Poisson's ratio (⅓ for aluminum)
$\quad t$ = plate thickness
$\quad b$ = width of plate

Equation (7.1) is equated to the Euler equation to obtain an equivalent slenderness ratio:

Euler equation: $\qquad\qquad \sigma = \dfrac{\pi^2 E}{\lambda_p^2}$ $\qquad\qquad$ (7.2)

where λ_p = equivalent slenderness ratio;

Equivalent slenderness ratio of plate:

$$\lambda_p = \sqrt{\frac{1 - \nu^2}{k}}\frac{b}{t} = C\frac{b}{t} \tag{7.3}$$

where C = buckling coefficient. The values from Eq. (7.3) are inserted into equations for elastic and inelastic buckling of plates. The buckling coefficients for many proportions of plate and loading conditions have been published in many places. Table 7.1 provides buckling coefficients of plates that are relatively long compared to the width.

7.1.2 Inelastic buckling of plates

Straight-line portions of curves representing inelastic buckling similar to those discussed previously for columns and beams, have been developed for aluminum plates.[1,3] The coefficients are presented in Table 7.2. Figure 7.2 shows that there is good agreement between calculated and test data for plates under uniform compression.

TABLE 7.1 Equivalent Slenderness Ratio λ_p, Long Plates[2,3]

Case					
	Support Conditions on Longitudinal Edges				
Uniform end loads	$5.13\dfrac{b}{t}$	$2.89\dfrac{b}{t}$	$1.63\dfrac{b}{t}$	$1.40\dfrac{b}{t}$	$1.24\dfrac{b}{t}$
Varying end loads	$4.45\dfrac{b}{t}$	$2.62\dfrac{b}{t}$	$1.17\dfrac{b}{t}$	$1.06\dfrac{b}{t}$	$0.89\dfrac{b}{t}$
Varying end loads	$3.64\dfrac{b}{t}$	$2.27\dfrac{b}{t}$	$0.67\dfrac{b}{t}$	$0.67\dfrac{b}{t}$	$0.52\dfrac{b}{t}$
Varying end loads	$2.56\dfrac{b}{t}$	$1.36\dfrac{b}{t}$	$1.17\dfrac{b}{t}$	$1.06\dfrac{b}{t}$	$0.89\dfrac{b}{t}$
Shear			$1.41\dfrac{b}{t}$	$1.09\dfrac{b}{t}$	

7.2 Postbuckling Behavior of Plates

7.2.1 Maximum strength

The axial stresses in a plate under uniform compression before and after buckling are shown schematically in Fig. 7.3. After buckling, the stresses become nonuniform, with high values at the edges compared to values at the interior of the plate. The average stress at failure for thin plates is much higher than the buckling stress. The average stress at maximum load is often referred to as the *crippling stress*. The mechanism for failure of plates with a stress gradient over the width (in-plane bending) is similar to that shown.[4]

A different mechanism occurs above the buckling value in shear panels as illustrated in Fig. 7.4. In this case, folds develop in the plate at a 45° angle. The folds are buckles caused by the compressive component of the shear forces. After buckling, the tension component in-

TABLE 7.2 Buckling Constants for Plates[1,3]

Loading on plate	Tempers	B_p (ksi)	B_p (MPa)	D_p (ksi)	D_p (MPa)	C_p
Axial compression	-O, -H, -T1, -T2, -T3, -T4	$\sigma_y\left[1 + \dfrac{(\sigma_y)^{1/3}}{7.6}\right]$	$\sigma_y\left[1 + \dfrac{(\sigma_y)^{1/3}}{14.5}\right]$	$\dfrac{B_p}{20}\left(\dfrac{6B_p}{E}\right)^{1/2}$	$\dfrac{B_p}{20}\left(\dfrac{6B_p}{E}\right)^{1/2}$	$\dfrac{2}{3}\dfrac{B_p}{D_p}$
	-T5, -T6, -T7, -T8, -T9	$\sigma_y\left[1 + \dfrac{(\sigma_y)^{1/3}}{11.4}\right]$	$\sigma_y\left[1 + \dfrac{(\sigma_y)^{1/3}}{21.7}\right]$	$\dfrac{B_p}{10}\left(\dfrac{B_p}{E}\right)^{1/2}$	$\dfrac{B_p}{10}\left(\dfrac{B_p}{E}\right)^{1/2}$	$0.41\dfrac{B_p}{D_p}$
Bending	All	$1.3\sigma_y\left[1 + \dfrac{(\sigma_y)^{1/3}}{7.0}\right]$	$1.3\sigma_y\left[1 + \dfrac{(\sigma_y)^{1/3}}{13.3}\right]$	$\dfrac{B_p}{20}\left(\dfrac{6B_p}{E}\right)^{1/2}$	$\dfrac{B_p}{20}\left(\dfrac{6B_p}{E}\right)^{1/2}$	$\dfrac{2}{3}\dfrac{B_p}{D_p}$
Shear	-O, -H, -T1, -T2, -T3, -T4	$\tau_y\left[1 + \dfrac{(\tau_y)^{1/3}}{6.2}\right]$	$\tau_y\left[1 + \dfrac{(\tau_y)^{1/3}}{11.8}\right]$	$\dfrac{B_p}{20}\left(\dfrac{6B_p}{E}\right)^{1/2}$	$\dfrac{B_p}{20}\left(\dfrac{6B_p}{E}\right)^{1/2}$	$\dfrac{2}{3}\dfrac{B_p}{D_p}$
	-T5, -T6, -T7, -T8, -T9	$\tau_y\left[1 + \dfrac{(\tau_y)^{1/3}}{9.3}\right]$	$\tau_y\left[1 + \dfrac{(\tau_y)^{1/3}}{17.7}\right]$	$\dfrac{B_p}{10}\left(\dfrac{B_p}{E}\right)^{1/2}$	$\dfrac{B_p}{10}\left(\dfrac{B_p}{E}\right)^{1/2}$	$0.41\dfrac{B_p}{D_p}$

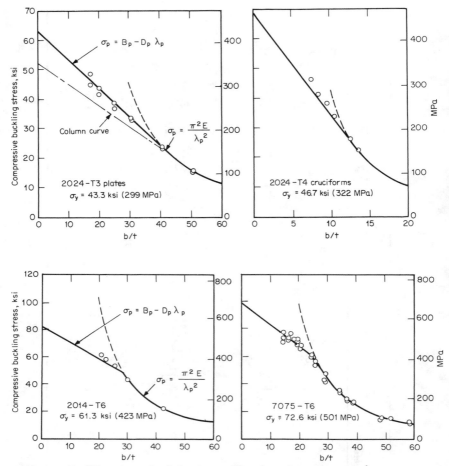

Figure 7.2 Buckling strength of aluminum alloy plates in compression.[1]

creases until failure occurs, while the compressive component stays essentially constant. It is common when designing to limit the stresses to the tensile yield strength of the material.

Equations for the postbuckling strength of plates in uniaxial compression and in-plane bending are given in Table 7.3. The equations have the following form:[5]

$$\sigma = \frac{K_2\sqrt{B_p E}}{\lambda_p} \tag{7.4}$$

where σ = crippling strength
B_p = coefficient for buckling in the inelastic range of the material

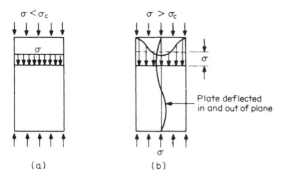

Figure 7.3 Stress distribution in plates under compression. (a) Prebuckled stresses, (b) postbuckled stresses.

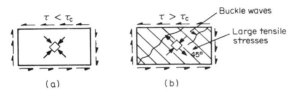

Figure 7.4 Stress distribution in plates under shear. (a) Prebuckled stresses, (b) postbuckled stresses.

E = modulus of elasticity
λ_p = equivalent slenderness ratio [Eq. (7.3) and Table 7.1]
K_2 = coefficient as defined in Table 7.3

Equation (7.4) applies if the equivalent slenderness is equal to or greater than the following value:

$$\lambda_p = \frac{K_1 B_p}{D_p} = C_p \tag{7.5}$$

where B_p, D_p = constants for inelastic buckling of plates
K_1 = constant defined in Table 7.3
C_p = slenderness ratio for intersection of elastic and inelastic curves

If the equivalent slenderness ratio is less than the value given by Eq. (7.5), the crippling stress is the same as the buckling stress calculated by the straight-line formula for inelastic buckling (see Sec. 7.1.2).

A simplified procedure has been suggested to calculate the crippling strength of sections in compression or bending.[5] The strength of each

TABLE 7.3   Formulas for Postbuckling Strength of Plates[1,3]

Loading on plate	Tempers	K_1 $$\lambda_p \geq \frac{K_1 B_p}{D_p} = C_p$$	K_2, stress for $\lambda_p > C_p$ $$\sigma = \frac{K_2 \sqrt{B_p E}}{\lambda_p}$$
Axial compression	-O, -H, -T1, -T2, -T3, -T4	0.50	2.04
	-T5, -T6, -T7, -T8, -T9	0.35	2.27
Bending	-O, -H, -T1, -T2, -T3, -T4	0.50	2.04
	-T5, -T6, -T7, -T8, -T9	0.50	2.04
Shear	All	$$\lambda_p \geq \frac{B_p - \tau_y}{D_p}$$	$\tau = \tau_c + 0.866(\tau_y - \tau_c)$

plate element is calculated using Eq. (7.4), assuming that the edges
are simply supported. A weighted-average stress of all elements is calculated. Failure is expected when the stress due to loading, calculated
in the usual manner by assuming that the entire section is effective,
reaches the weighted-average failure stress. For a trapezoidal section
in bending of uniform thickness t, with a compression flange of width
b, and an inclined web of width w, under bending the average failure
stress is as follows:

$$\sigma = \frac{bt\sigma_b + \frac{1}{3}wt\sigma_w}{bt + \frac{1}{3}wt} \tag{7.6}$$

where σ_b = failure stress of flange (plates in uniform compression)
   σ_w = failure stress of webs (plates in in-plane bending)
   b = width of compression flange
   t = thickness
   w = length of inclined web

In deriving Eq. (7.6), the compression flange is assumed to be the material further than two-thirds of the distance from the neutral axis to the extreme compression fiber. Figure 7.5 shows good agreement between calculated and test values using this method of calculation for thin members in bending.

Good agreement between test and calculations is also obtained for sections under uniform compression as illustrated in Fig. 7.6. Table 7.3 also provides the equation for the postbuckling strength of panels in shear.[6] The equation is as follows:

$$\tau = \tau_c + 0.866\,(\tau_y - \tau_c) \qquad (7.7)$$

where τ = postbuckling strength in shear
τ_c = calculated shear buckling stress
τ_y = shear yield strength

Equation (7.7) applies when the equivalent slenderness ratio λ_p for the panel is as follows:

$$\lambda_p \geq \frac{B_p - \tau_y}{D_p} \qquad (7.8)$$

where D_p, B_p = coefficients for shear buckling of plates (Table 7.2).

Figure 7.7 shows the curves for buckling and ultimate strength for a panel of 6061-T6. Considerable additional strength above that for buckling exists for thin panels. Figure 7.8 shows that the equation for the postbuckling strength of plates in shear, as given by Eq. (7.7) is generally conservative compared to test data. Part of the conservativeness is due to the use of yielding under combined stress as the criterion for failure, rather than a fracture criterion. The conservative assumption is helpful for some of the thin webs shown in Fig. 7.8, with low ratios of buckling to ultimate strength.

7.2.2 Effects of welds

Figure 7.9 shows ultimate compressive strength data from tests of aluminum plates,[7] with and without welds, and comparisons with calculated values. The upper curves are calculated using the coefficients provided in Tables 7.2 and 7.3 and the reported yield strengths for the base metals. The lower curve is based on all-welded properties. An intermediate curve shown for 6082-TF is calculated using an estimated 10-in (254-mm) gage length yield strength. Some observations of the results follow. The calculation procedure for the ultimate strength of nonwelded plates provided in this chapter and used as the basis for The Aluminum Association specifications[3] is satisfactory. The calculated values are in close agreement with test data for low b/t ratios

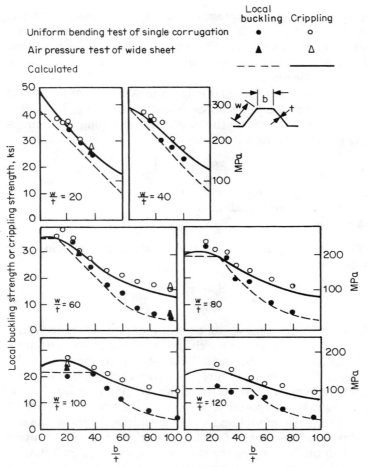

Figure 7.5 Strength of formed sheet sections in bending.[5]

and are conservative (up to 25 percent) for large b/t ratios. Welds decreased the strength compared to unwelded plates but the decrease was not as much as that calculated for all-welded material. Calculations based on a 10-in (254-mm) gage length yield strength were closer to the test data and conservative. It appears that the yield strength of the longer gage length may be used for the design of welded plates, but better design techniques need to be developed.

7.3 Buckling and Postbuckling Behavior of Plates with Partial Edge Loads

The ability of a thin-walled member to accommodate a concentrated load, usually referred to as high *web-crippling strength*, is a requisite

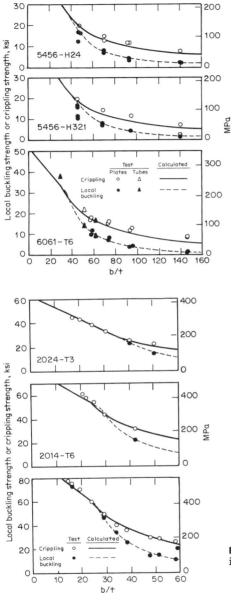

Figure 7.6 Strength of flat plates in compression.[5]

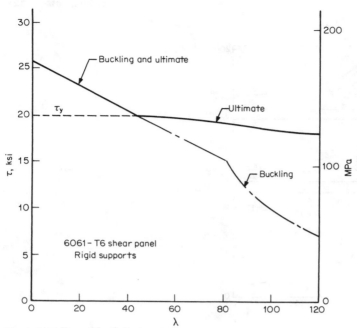

Figure 7.7 Strength of shear panels.

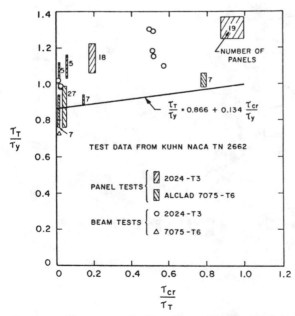

Figure 7.8 Shear strength of panels and beams with rigid flanges.[6]

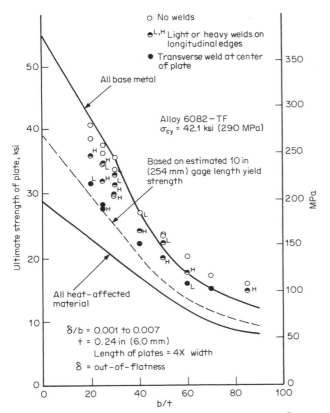

Figure 7.9 Postbuckling strength of plates with welds.[7]

for many products in the building and transportation markets. Thin-walled ribbed building products subjected to foot traffic loads or reactions at supports, and automotive bumpers under impact loads are examples. Although solutions for buckling have been developed for several cases, most practical products have built-in eccentricities and out-of-flatness so that the buckling value does not occur nor is it important for design.[8] Figure 7.10 is a schematic plot of the load applied versus the lateral deflection of a point under the load, but partway down the depth of the plate.[9] The figure looks similar to that given for plates under compression; however, the mechanism for load carrying above buckling is much different. In this case, the additional load carrying above buckling results from a larger area of the web involved in carrying the load. The entire depth of the web is bowed laterally. At ultimate load, a crescent-shaped yield line forms as shown in Fig. 7.11. In thin webs, a somewhat higher load may develop with large

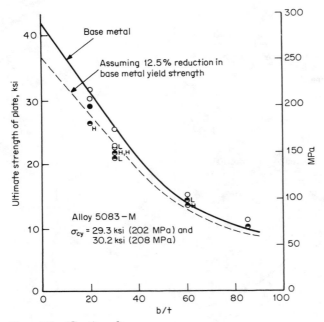

Figure 7.9 *(Continued)*

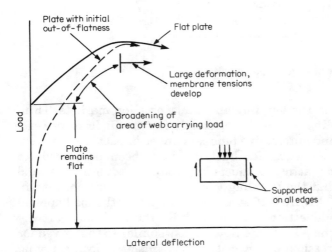

Figure 7.10 Behavior of plate under partial edge loading.

Figure 7.11   Failure under concentrated load.

vertical displacements because of membrane tensile forces that develop under the load points in the longitudinal direction.

Figure 7.12 shows the test setup chosen as the basis for determining web-crippling strength. The flange opposite that loaded is essentially fixed against deflection and rotation. This setup prevents any longitu-

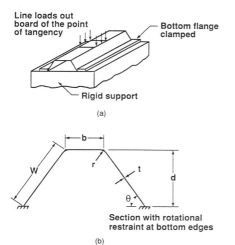

Figure 7.12   Test setup (a) and section dimensions (b).

dinal bending in the section and develops the largest concentrated load.[10,11] If the member has bending as well as concentrated loading, the combined effects must be considered (see Sec. 7.4). Tests and calculations show that the maximum load acting on one web is the following:

$$P = \frac{t^2(N + C_1)(\sin\theta)\left(0.46\sigma_y + 0.02\sqrt{E\sigma_y}\right)}{C_2 + r(1 - \cos\theta)} \tag{7.9}$$

where P = maximum load on one web
C_1 = 5.4 in or 137 mm, consistent with other units used
C_2 = 0.4 in or 10 mm, consistent with other units used
N = length of load
σ_y = compressive yield strength
E = modulus of elasticity
r = radius between web and top flange
t = thickness

The definition of terms for a trapezoidal section is shown in Fig. 7.12. Values from Eq. (7.9) are conservative compared to test and simulation results using the computer program DYNA3D as shown in Fig. 7.13. Trapezoidal sections, rectangular box sections, and I-sections are included in the data. A similar equation gives reasonable values when the load is applied to the end of the member:

$$P_e = \frac{1.2t^2(N + C_{1e})(\sin\theta)\left(0.46\sigma_y + 0.02\sqrt{E\sigma_y}\right)}{C_2 + r(1 - \cos\theta)} \tag{7.10}$$

where C_{1e} = 1.3 in or 33 mm, consistent with other units used
C_2 = 0.4 in or 10 mm, consistent with other units used

The other terms are as identified above.

There is good agreement between calculated values using Eq. (7.10) and test and simulation results as noted on Fig. 7.14. Most practical structures with concentrated loads on thin-walled members are beams. In these cases there is combined bending and concentrated load, the subject covered in the next section.

7.4 Interaction of Various Types of Buckling and Crippling

7.4.1 Combined bending and web crippling

Figure 7.15 shows the distortions that occur in a thin-walled beam under a concentrated load. The vertical deflections lower the strength of

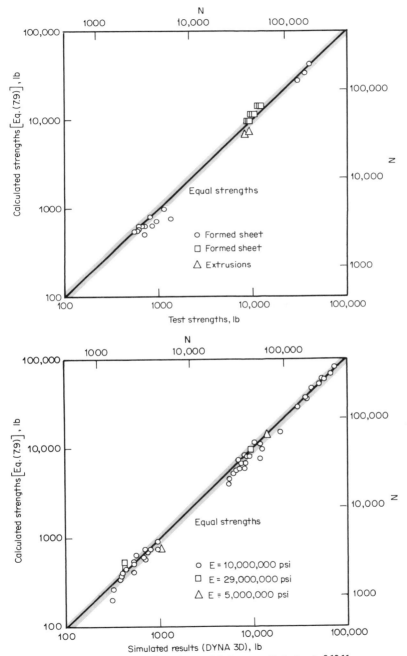

Figure 7.13 Web crippling of sections with intermediate loads.[8,10,11]

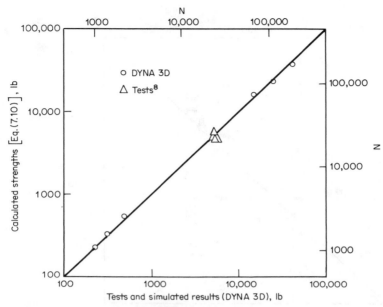

Figure 7.14 Web crippling of sections with end loads.

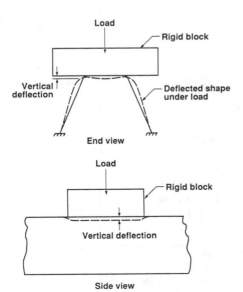

Figure 7.15 Distortions of a thin-walled part under concentrated load.

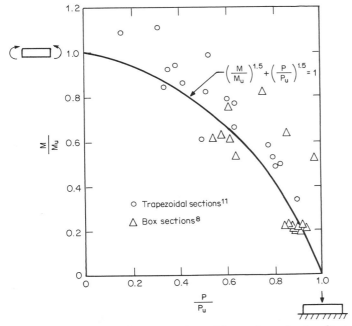

Figure 7.16 Interaction between web crushing and member bending.

the section acting as a beam. Thus, the two modes of failure interact as shown in Fig. 7.16. Data from tests of trapezoidal and box sections are shown. The interaction formula is as follows:

$$\left(\frac{M}{M_u}\right)^{1.5} + \left(\frac{P}{P_u}\right)^{1.5} \le 1 \qquad (7.11)$$

where M = applied moment
$\quad P$ = applied concentrated load
$\quad M_u$ = maximum moment (pure bending, see Sec. 7.2)
$\quad P_u$ = maximum load, see Eq. (7.9)

Equation (7.11) is reasonable for design.

7.4.2 Buckling under axial, bending, and shear stresses

Aluminum design specifications[3] allow for the effects of some combined loading on flat plates (see Fig. 7.17) by the following interaction equation (note that the specifications are expressed in terms of allowable stress, whereas Eq. (7.12) is written in terms of buckling stress):

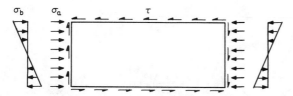

Figure 7.17 Plate under combined stresses.

$$\left(\frac{\sigma_a}{\sigma_{ab}}\right) + \left(\frac{\sigma_b}{\sigma_{bb}}\right)^2 + \left(\frac{\tau}{\tau_b}\right)^2 = 1 \qquad (7.12)$$

where σ_a = axial stress

σ_{ab} = buckling stress under axial loading

σ_b = bending stress

σ_{bb} = buckling stress under bending loading

τ = shear stress

τ_b = buckling stress under shear loading

The development of Eq. (7.12) is provided in the literature.[12] This equation is for buckling of plates, not ultimate strength. There is no available corresponding equation for ultimate strength.

7.5   Interaction of Local and Overall Buckling

The equations for column buckling and lateral buckling of beams presented in previous chapters apply only to those cases in which the elements of the section do not buckle. The previous sections of this chapter cover the cases in which elements buckle and fail without overall buckling of the component. If elements of the section buckle (local buckling) at a lower stress than that corresponding to column or beam buckling as defined in previous sections, the stiffness and strength of the component are reduced. The member buckles at a stress intermediate to that for column and local buckling. This intermediate strength is given by the following equation:[13]

$$\sigma_u = \sigma_E\left(\frac{\sigma_c}{\sigma_E}\right)^{2/3} \qquad (7.13)$$

where σ_u = ultimate strength

σ_c = local buckling strength

$\sigma_E = \pi^2 E/\lambda^2$

λ = equivalent or effective slenderness ratio

E = modulus of elasticity

Figure 7.18 provides a comparison of calculations using Eq. (7.13) and test data for columns. Note that at stresses below those corresponding to local buckling, the normal-column curve applies. The short-column strength is conservatively given by the crippling strength. Equation (7.13) adequately predicts the data for the intermediate cases. Although Eq. (7.13) is empirical, it appears to apply well to many cases. Figure 7.19 shows comparisons for the combined local and lateral buckling of beams. It appears that a few of the T-

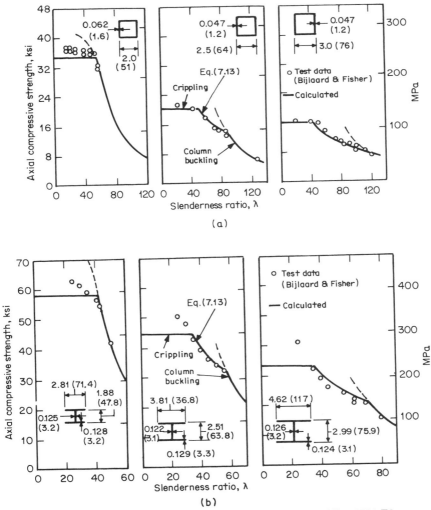

Figure 7.18   Combined local and overall buckling of columns. (a) Alloy 6061-T6: square tubes. (b) Alloy 7075-T6: H-sections.[13]

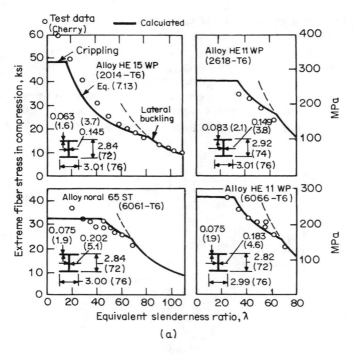

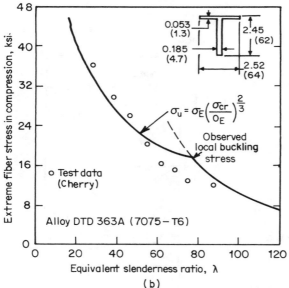

Figure 7.19   Combined local and lateral buckling. (a) I-sections, (b) T-sections.[13]

section beams may have buckled locally below the observed stress that was listed with the test results. Otherwise, the agreement between calculated and test values is good. Note also that the columns failed by flexural buckling only. Thus, Eq. (7.13) may or may not apply to other shapes or failure modes.

References

1. Clark, John W., and Rolf, Richard L., "Buckling of Aluminum Columns, Plates, and Beams," *Proceedings of the American Society of Civil Engineers,* Journal of the Structural Division, June 1966.
2. *Alcoa Structural Handbook,* Aluminum Company of America, Pittsburgh, 1960.
3. "Specifications for Aluminum Structures," *Aluminum Construction Manual,* 4th ed., The Aluminum Association, Washington, D.C., 1986.
4. Jombock, J. R., and Clark, J. W., "Post Buckling Behavior of Flat Plates," *Proceedings of the American Society of Civil Engineers,* Journal of the Structural Division, June 1961.
5. Jombock, John R., and Clark, John W., "Bending Strength of Aluminum Formed Sheet Members," *Proceedings of the American Society of Civil Engineers,* February 1968.
6. Sharp, Maurice L., and Clark, John W., "Thin Aluminum Shear Webs," *Proceedings of the American Society of Civil Engineers,* Journal of the Structural Division, April 1971.
7. Mofflin, D. S., and Dwight, J. B., "Tests on Individual Aluminum Plates under In-Plane Compression," University of Cambridge, Department of Engineering, Technical Report No. CVED/D-Struct/TR100, 1983.
8. Sharp, Maurice L., "Aluminum Beam Webs under Concentrated Loads," *Proceedings of the Specialty Conference, Methods of Structural Analysis,* University of Wisconsin–Madison, August 22–25, 1976.
9. Sharp, Maurice L., "Behavior of Plates under Partial Edge Loading," *Steel Structures Proceedings,* American Society of Civil Engineers, Structures Congress '89, San Francisco, May 1–5, 1989.
10. Sharp, Maurice L., and Jaworski, Andrzej P., "Methodology for Developing an Engineering Solution for Web Crippling," paper delivered at American Society of Civil Engineers Engineering Mechanics Specialty Conference, Columbus, Ohio, May 19–22, 1991.
11. Sharp, Maurice L., "Experimental and Theoretical Results for Web Crippling of Aluminum Trapezoidal Sections," in *Proceedings of the International Conference on Steel and Aluminum Structures, Singapore,* May 1991, Elsevier Applied Science Ltd., Essex, England, 1991.
12. Galambos, Theodore V. (ed.), *Guide to Stability Design Criteria for Metal Structures,* 4th ed., Wiley, New York, 1988.
13. Sharp, Maurice L., "Strength of Beams or Columns with Buckled Elements," *Proceedings of the American Society of Civil Engineers,* Journal of the Structural Division, Technical Notes, May 1970.

8

Stiffened Flat Plates

The efficiency of thin, wide, flat panels under compression or shear in the plane of the panel is improved considerably by the use of intermediate stiffeners. Failure of the stiffened components considered here is initiated by buckling, and the stiffeners increase both the buckling and ultimate strength. Information is presented for five types of components: (1) stiffening lips on outstanding flanges (long plates supported at one longitudinal edge) under uniform compression, (2) longitudinal stiffeners on plates supported on both longitudinal edges and loaded by uniform compression, (3) elastically supported compression flanges of members loaded in bending, (4) girders with transverse stiffeners loaded by shear, and (5) corrugated plates loaded by shear.

The behavior of these components is more complex than that for the elements covered previously, in that there are more failure modes to consider. There is generally a smaller experimental data base available compared to those for the problems previously covered; thus, the designer needs to exercise more judgment and conservatism in the development of the final configuration. Fortunately, commercial computer software is available that provides the designer the option of doing simulations of critical components to arrive at a more precise understanding of the behavior of the part.

8.1 Stiffening Lips on Flanges under Uniform Compression

Two buckling modes are illustrated for the hat-shaped section under uniform compression illustrated in Fig. 8.1: torsional buckling of the lipped flange and local buckling of the cross section. The component is assumed to be short enough so that it does not fail by overall flexural, torsional, or flexural-torsional buckling. Local buckling has been discussed in previous chapters. In torsional buckling the combined lip

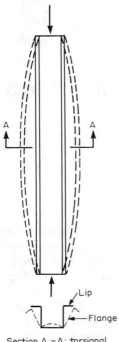

Section A – A: torsional
buckling of lipped flange

Local buckling of
section

Figure 8.1 Buckling of lipped flange under uniform compression.

and flange twist about an enforced axis of rotation at the base of the flange at failure. The half-wave of buckle is generally long compared to the width of the flange, and depends on the size of the lip and the proportions of the section.

An equivalent slenderness ratio has been defined[1] assuming that the lip and flange behave as a shape buckling in torsion about an enforced axis of rotation with an elastic restraint against rotation. The equation is as follows for the case of loaded edges that are free to rotate:

$$\lambda = \pi \sqrt{\frac{I_p}{\frac{3}{8}J + 2\sqrt{C_w(K_\phi/E)}}} \tag{8.1}$$

where λ = equivalent slenderness ratio

$I_p = I_{xo} + I_{yo}$ = polar moment of inertia of lip and flange about center of rotation

I_{xo}, I_{yo} = moments of inertia of lip and flange about the horizontal and vertical axes at the center of rotation

K_ϕ = elastic restraint factor

J = torsion constant of flange and lip

$C_w = b^2[I_{yc} - (bt^3/12)]$, warping term for lipped flange about center of rotation

I_{yc} = moment of inertia of flange and lip about their centroidal axis

E = modulus of elasticity

The nomenclature for the various terms is given in Fig. 8.2. The restraint against rotation depends on the proportions of the component and can be estimated from an analysis of a unit length of the shape. For the hat-shaped component given in Fig. 8.3 the restraint is the following:[1]

$$K_\phi = \frac{2D_w}{b_w}\left(\frac{1}{1 + \dfrac{2}{3}\dfrac{D_w}{D}\dfrac{h}{b_w}}\right) \tag{8.2}$$

where K_ϕ = torsional restraint
$D_w = Et_w^3/12(1 - v^2)$
$D = Et^3/12(1 - v^2)$
b_w = width of web of hat-shaped section
v = Poisson's ratio (1/3)
h = distance to centroid of combined lip and flange

Equation (8.2) gives the torsional restraint against rotation as calculated from the application of unit outward forces at the centroid of the combined lip and flange. The torsional restraint for other types of components would be different but could be estimated in a similar manner. Tests also could be used for cases in which calculations may not be very accurate (section contains joints, for example).

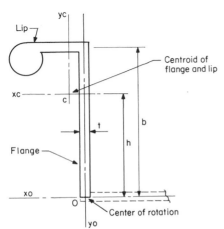

Figure 8.2 Nomenclature for buckling of lipped flange.

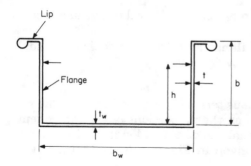

Figure 8.3   Hat-shaped section.

A comparison of test and calculations is provided in Fig. 8.4.[1] Calculations for fixed loaded edges are also shown because the length of the specimens tested was relatively short and affected the results. The fixed-edge case is developed in Ref. 1, but it requires an iterative solution and it would not be needed for long members. Straight-line formulas for column buckling were employed for all cases although plate buckling formulas are probably better for cases with small lips. Local buckling did not occur in the sections.

Figure 8.5 shows another set of data; here, local buckling did occur in most cases, sometimes preceding torsional buckling, sometimes after torsional buckling. The agreement with calculated values is satisfactory. There is insufficient data to define the interaction between local and overall buckling, but there likely is one, probably similar to that presented previously for columns and beams. Test information used to select local and torsional buckling loads for these shapes is presented in Fig. 8.6. The maximum loads carried by the shapes were higher than the buckling value selected for torsional buckling, suggesting that there is postbuckling strengthening.

Figure 8.7 shows the results of these tests and results as calculated using Eq. (8.1). The calculation procedure is conservative, but the designer should evaluate each new case for both local and torsional buckling modes.

8.2   Longitudinal Stiffeners under Compression

An equivalent slenderness ratio for buckling of a longitudinally stiffened plate has been proposed as follows:[1]

$$\lambda = \frac{4}{\sqrt{3}} \frac{b}{t} N \sqrt{\frac{1 + A_s/bt}{1 + \sqrt{32/3(I_e/t^3 b) + 1}}} \tag{8.3}$$

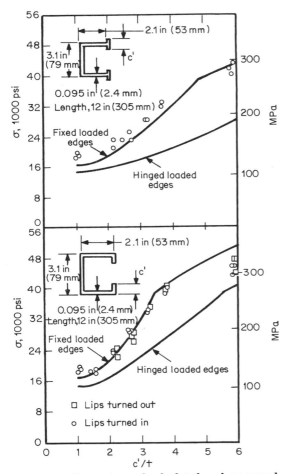

Figure 8.4 Comparison of calculated and measured buckling strengths of lipped flanges.[1]

where λ = equivalent slenderness ratio

A_s = area of stiffener (see Fig. 8.8)

I_e = moment of inertia of section consisting of stiffener plus width of plate equal to the stiffener spacing (see Fig. 8.8)

b = stiffener spacing

t = thickness of plate

N = number of panels in which the longitudinal stiffeners subdivide the plate

Figure 8.8 shows cross sections of longitudinally stiffened plates in which the stiffener is unsymmetrical with respect to the plate, and

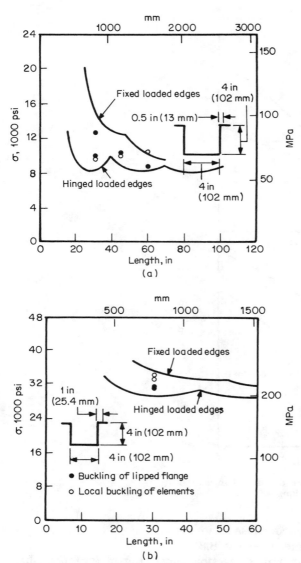

Figure 8.5 Buckling of lipped flanges of formed sections. (a) 0.064-in-thick (1.6-mm) 6061-T6. (b) 0.126-in-thick (3.2-mm) 6061-T6.[1]

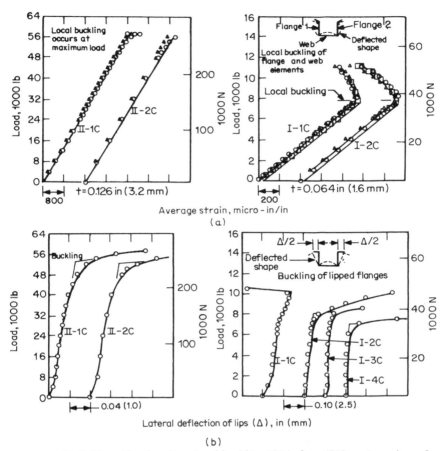

Figure 8.6 Definition of local and torsional buckling [30-in-long (762-mm) specimens]. (a) Load-average strain curves, (b) load-deflection curves for lipped flanges.[1]

provides a definition of the stiffener section. In the case of a stiffener that is symmetrical about the plate, the plate would not be used as part of the stiffener section. More than one stiffener could be used to stiffen the plate.

Equation (8.3) can be used for stiffeners on plates in axial compression or on plates that are acting as the compression flanges of beams. Figures 8.9, 8.10, and 8.11 compare the results of calculations and bending tests on formed-sheet beams of various proportions with stiffeners formed into the compression flanges. All the test specimens were loaded in uniform bending. For each type of beam the depth of the stiffener y was varied. The failure stresses for the test beams were calculated from the maximum moment divided by the section modulus.

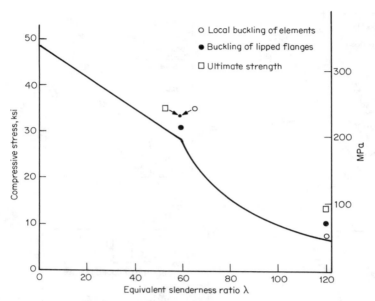

Figure 8.7 Buckling/ultimate strength of lipped flanges.

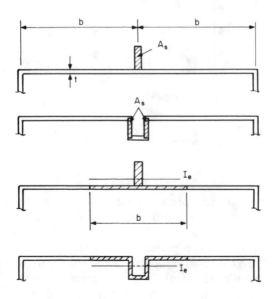

Figure 8.8 Longitudinal stiffener. ($N = 2$ for all.)

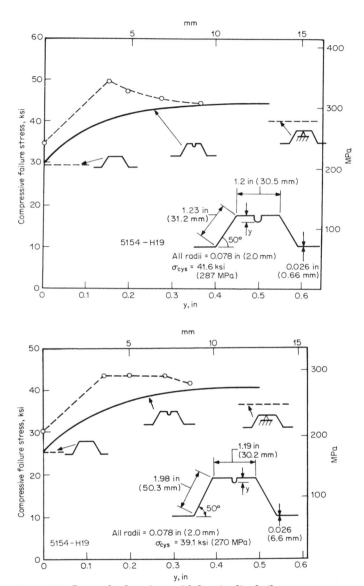

Figure 8.9 Strength of sections with longitudinal ribs.

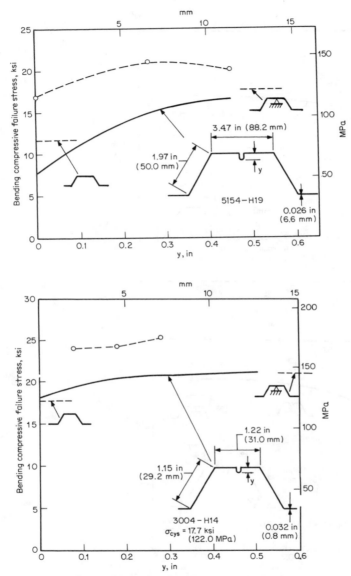

Figure 8.10 Strength of sections with one longitudinal rib.

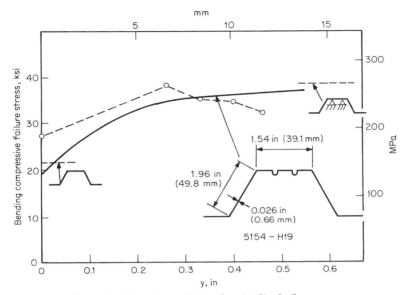

Figure 8.11 Strength of sections with two longitudinal ribs.

The weighted-average procedure described in Sec. 7.2.1 for the calculation of average stress at failure was employed. For the flange, the
buckling stress as calculated by Eq. (8.3) and the entire area (including stiffener) were employed. The straight-line formula for plates
(rather than the column curve) was used in the inelastic region because it provided a better fit with the data. The crippling strength of
the webs was used in the weighted-average calculations.

Two other reference calculations were made and shown in the figures; the calculated crippling strength of the section without a rib
(left-side dashed line) and the calculated crippling strength of the section with a line support at the location of the rib (right-side dashed
line).

Several observations are as follows:

- The calculation procedure described above was conservative compared to test data—in some cases very conservative.

- Longitudinal ribs improved the strength of the sections considerably
(one-stiffener case).

- Relatively small ribs (depth = ⅛ flange width) were effective.
Larger ribs were no more effective (one-stiffener case).

- There was no attempt to take into account possible strengthening
effects of radii or any postbuckling behavior of the stiffened panel.

The difference in calculated values on Fig. 8.10 for $y = 0$ occurs because postbuckling strength was used for plate calculations, but buckling only was considered when using Eq. (8.3).

Based on limited data and calculations, relatively small ribs can improve the efficiency of plates under uniform compression.

8.3 Elastically Supported Compression Flanges

Figure 8.12 shows some practical examples of products of this type; the members are loaded in bending and the combined flange and web may fail by torsional buckling about the intersection of the web and tensile element, with elastic restraint against torsion provided by the tensile flange elements.

Figure 8.13 shows a hat-shaped specimen under test with loads being applied at the quarter points of the span. This problem is similar to that discussed earlier in Sec. 8.1, but there are some differences that need to be considered. Similarities and differences are listed here:

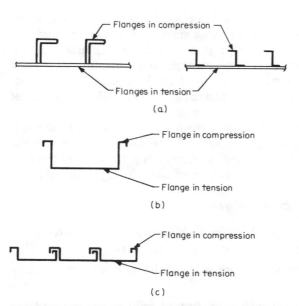

Figure 8.12 Structural members with unsupported compression flanges. (a) Stiffened sheet and plate in bending (ship construction, stiffeners on sign panels, etc.). (b) Trough-type cable trays. (c) Raised-seam roofing.

Figure 8.13   Test of hat-shaped section.

- The lipped flange considered previously is under uniform compression, while the unsupported compression flange is under bending (has a stress gradient that varies from compression to tension).

- Failure in both cases can be by torsion about an enforced axis of rotation or by local buckling of individual elements.

- The elastic restraint offered to the unsupported compression flange is relatively higher than that for lipped flanges under uniform compression because the supporting elements are in tension rather than compression.

- All unsupported compression flanges that are not symmetrical about the web have lateral bending moments because of the shear transfer between web and flange, whereas lipped flanges have none.

Figure 8.14 describes conditions that cause lateral bending in unsupported compression flanges as mentioned in the last item above. Symmetrical flanges about the web shown on the left have shear transfer forces that occur at the neutral axis of the flange section, and lateral bending moments are not present for perfectly straight flanges. The unsymmetrical flanges shown on the right do have lateral bending moments because the shear transfer forces are eccentric

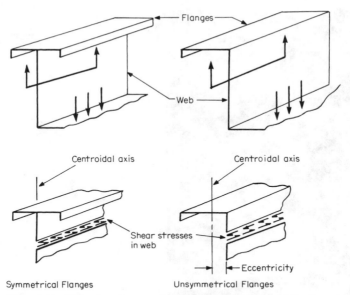

Figure 8.14 Behavior of symmetrical and unsymmetrical flanges.

relative to the neutral axis of the flange. Note that all of the cases shown in Fig. 8.12 are of this type. The practical consequence of the eccentric loading in this case is that the flanges deflect laterally as soon as load is applied. The lateral deflection reduces the vertical stiffness of the shape (vertical deflections are higher than that calculated using the original shape), and the buckling strength may be affected. Figure 8.15 shows some test data and some simulations (using computer code DYNA3D) of sections with symmetrical and unsymmetrical flanges. The simulations provide a reasonable estimate of the load–lateral deflection curves. Also, the curves for the symmetrical and unsymmetrical cases, while different, seem to approach a similar value at large deformations.

One possibility for preliminary design is to use the solution for torsional buckling provided previously for lipped flanges. That equation obviously is approximate for this case because it does not explicitly take into account all the behavior mentioned above. The equivalent slenderness is as follows:

$$\lambda = \pi \sqrt{\frac{I_p}{\frac{3}{8} J + 2\sqrt{C_w(K_\phi/E)}}} \tag{8.4}$$

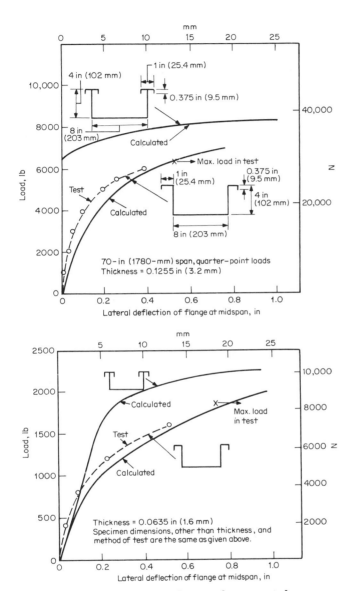

Figure 8.15 Load vs. lateral deflection of unsupported compression flanges.

Equation (8.4) is exactly the same as Eq. (8.1) and the definitions of terms are the same.

Table 8.1 provides some test data for formed-sheet beam sections to compare against calculated values. Specimens from two alloys, several span lengths, and different restraints at the ends are included. The types of specimens are defined in Fig. 8.16.

Figures 8.17 and 8.18 show the comparison of calculated results using Eq. (8.4) and column-buckling curves for the inelastic range, and experimental results from Table 8.1. The agreement between calculated and test data is only fair but there are many effects neglected in the analysis. Some other observations may be of interest. Torsional buckling is important to consider. Member length and restraint at the ends were not very important. The effect of eccentricity of loading is compensated for in the conservativeness of the solution used. Although these specimens did not buckle locally prior to failure, the appearance after failure was similar to that from crippling of the elements (see Fig. 8.19).

Figure 8.20 presents the result of a test of a stiffened panel for ship construction and a comparison with calculations based on Eq. (8.4).

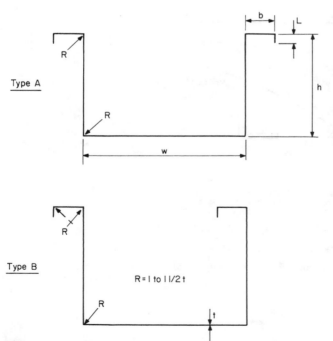

Figure 8.16 Sections used to evaluate unsupported compression flanges.

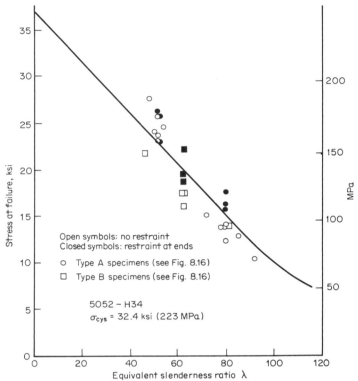

Figure 8.17 Failure stresses for formed sheet beams with unsupported compression flanges.

The agreement is reasonable. Large lateral deflections of the flange occurred at ultimate load.

The calculation method presented here is satisfactory for preliminary design for strength. The vertical deflection will be significantly greater than that calculated using original section properties. If better accuracy is needed, either (1) computer simulations using large-deflection, inelastic analysis or (2) tests may be employed.

8.4 Girder Webs

The design of thin-web, stiffened girders is discussed in this section.[2] Common current uses for these girders are the side walls of trailers for trucks, sides of railway cars, and parts of aircraft. Figure 8.21 shows a girder after a tensile failure of the web plate. Diagonal tension folds have developed and extend over stiffeners that are too small to contain the buckles to the flat panels. The design of the girder must take into ac-

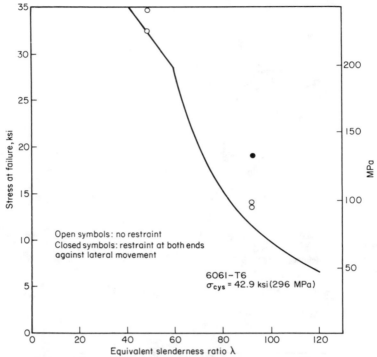

Figure 8.18 Failure stresses for formed sheet beams with unsupported compression flanges.

count the stresses in the web, and the effects of the diagonal tension forces on the fastenings, the intermediate stiffeners, and the flanges. The effects are dependent on the bending rigidity of the flanges.

8.4.1 Analysis of the girder web

A simple analysis has been proposed that assumes that the web in the postbuckled condition has a combined stress consisting of portions of those given in Fig. 8.22.[2] The maximum shear stress causing yielding in the web is the following:

$$\tau_T = \tau_{cr} + C_1 \frac{\sqrt{3}}{2} (\tau_y - \tau_{cr}) \tag{8.5}$$

where τ_T = shear stress causing yielding in the web
$\quad C_1$ = coefficient (Fig. 8.23)
$\quad \tau_{cr}$ = shear buckling stress
$\quad \tau_y$ = shear yield strength

TABLE 8.1 Bending Strength of Formed Sections with Unsupported Compression Flanges*

Alloy	Lateral restraint	Span, in (mm)	t, in (mm)	b, in (mm)	L, in (mm)	h, in (mm)	W, in (mm)	Type	Calculated stress at maximum load, ksi (MPa)
5052-H34	None	70 (1780)	0.0635 (1.61)	0.5 (12.7)	0 (0)	4.0 (102)	4.0 (102)	A	10.2 (70)
6061-T6	None	34 (860)	0.0625 (1.59)	0.5 (12.7)	0 (0)	4.0 (102)	4.0 (102)	A	13.9 (96)
6061-T6	None	70 (1780)	0.0625 (1.59)	0.5 (12.7)	0 (0)	4.0 (102)	4.0 (102)	A	14.2 (98)
6061-T6	Both ends	34 (860)	0.0625 (1.59)	0.5 (12.7)	0 (0)	4.0 (102)	4.0 (102)	A	19.3 (133)
6061-T6	Both ends	70 (1780)	0.0625 (1.59)	0.5 (12.7)	0 (0)	4.0 (102)	4.0 (102)	A	14.6 (101)
5052-H34	None	70 (1780)	0.0635 (1.61)	0.75 (19.0)	0.25 (6.4)	4.0 (102)	4.0 (102)	A	13.7 (94)
5052-H34	None	70 (1780)	0.635 (1.61)	1.0 (25.4)	0.375 (9.5)	4.0 (102)	4.0 (102)	A	15.1 (104)
5052-H34	None	46 (1170)	0.0635 (1.61)	1.0 (25.4)	0.375 (9.5)	4.0 (102)	8.0 (203)	A	12.3 (85)
5052-H34	Both ends	46 (1170)	0.0635 (1.61)	1.0 (25.4)	0.375 (9.5)	4.0 (102)	8.0 (203)	A	15.8 (109)
5052-H34	None	70 (1780)	0.0635 (1.61)	1.0 (25.4)	0.375 (9.5)	4.0 (102)	8.0 (203)	A	14.0 (97)
5052-H34	Both ends	70 (1780)	0.0635 (1.61)	1.0 (25.4)	0.375 (9.5)	4.0 (102)	8.0 (203)	A	17.6 (121)
5052-H34	None	118 (3000)	0.0635 (1.61)	1.0 (25.4)	0.375 (9.5)	4.0 (102)	8.0 (203)	A	13.9 (96)

NOTE: See page 155 for footnote.

TABLE 8.1 Bending Strength of Formed Sections with Unsupported Compression Flanges* (Continued)

Alloy	Lateral restraint	Span, in (mm)	t, in (mm)	b, in (mm)	L, in (mm)	h, in (mm)	W, in (mm)	Type	Calculated stress at maximum load, ksi (MPa)
5052-H34	Both ends	118 (3000)	0.0635 (1.61)	1.0 (25.4)	0.375 (9.5)	4.0 (102)	8.0 (203)	A	16.3 (112)
5052-H34	None	70 (1780)	0.0635 (1.61)	1.0 (25.4)	0.375 (9.5)	4.0 (102)	12.0 (305)	A	12.8 (88)
5052-H34	None	70 (1780)	0.0635 (1.61)	1.0 (25.4)	0.375 (9.5)	4.0 (102)	8.0 (203)	B	14.0 (97)
5052-H34	None	46 (1170)	0.0635 (1.61)	1.0 (25.4)	0.375 (9.5)	3.0 (76)	8.0 (203)	B	16.1 (111)
5052-H34	Both ends	46 (1170)	0.0635 (1.61)	1.0 (25.4)	0.375 (9.5)	3.0 (76)	8.0 (203)	B	19.6 (135)
5052-H34	None	70 (1780)	0.0635 (1.61)	1.0 (25.4)	0.375 (9.5)	3.0 (76)	8.0 (203)	B	17.6 (121)
5052-H34	Both ends	70 (1780)	0.0635 (1.61)	1.0 (25.4)	0.375 (9.5)	3.0 (76)	8.0 (203)	B	22.3 (154)
5052-H34	None	118 (3000)	0.0635 (1.61)	1.0 (25.4)	0.375 (9.5)	3.0 (76)	8.0 (203)	B	17.3 (119)
5052-H34	Both ends	118 (3000)	0.0635 (1.61)	1.0 (25.4)	0.375 (9.5)	3.0 (76)	8.0 (203)	B	18.8 (130)
5052-H34	None	70 (1780)	0.0635 (1.61)	1.0 (25.4)	0.375 (9.5)	2.0 (51)	8.0 (203)	B	21.9 (151)
5052-H34	None	70 (1780)	0.1255 (3.19)	1.0 (25.4)	0 (0)	4.0 (102)	4.0 (102)	A	27.5 (190)
6061-T6	None	34 (860)	0.125 (3.18)	1.0 (25.4)	0 (0)	4.0 (102)	4.0 (102)	A	34.8 (240)

6061-T6	None	70 (1780)	0.125 (3.18)	1.0 (25.4)	0 (0)	4.0 (102)	4.0 (102)	A	32.5 (224)
5052-H34	None	46 (1170)	0.1255 (3.19)	1.0 (25.4)	0 (0)	4.0 (102)	8.0 (203)	A	23.6 (163)
5052-H34	Both ends	46 (1170)	0.1255 (3.19)	1.0 (25.4)	0 (0)	4.0 (102)	8.0 (203)	A	26.3 (181)
5052-H34	None	70 (1780)	0.1255 (3.19)	1.0 (25.4)	0 (0)	4.0 (102)	8.0 (203)	A	25.9 (179)
5052-H34	Both ends	70 (1780)	0.1255 (3.19)	1.0 (25.4)	0 (0)	4.0 (102)	8.0 (203)	A	25.9 (179)
5052-H34	None	118 (3000)	0.1255 (3.19)	1.0 (25.4)	0 (0)	4.0 (102)	8.0 (203)	A	23.1 (159)
5052-H34	Both ends	118 (3000)	0.1255 (3.19)	1.0 (25.4)	0 (0)	4.0 (102)	8.0 (203)	A	23.1 (159)
5052-H34	None	70 (1780)	0.1255 (3.19)	1.0 (25.4)	0.375 (9.5)	4.0 (102)	8.0 (203)	A	24.2 (167)
5052-H34	None	70 (1780)	0.1255 (3.19)	1.0 (25.4)	0 (0)	4.0 (102)	12.0 (305)	A	24.8 (171)

Specimens were 2 in (51 mm) longer than span.
Load was applied over areas ½ in (12.7 mm) wide by 2 in (51 mm) long, adjacent to webs and at quarter points of span.
*See Fig. 8.16 for definitions of terms.

Figure 8.19 Failure of unsupported compression flanges.

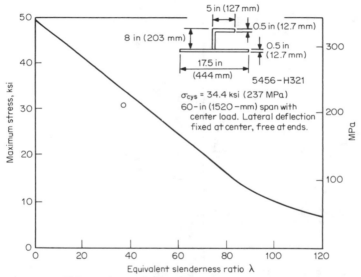

Figure 8.20 Ship section under bending.

The stress as given by Eq. (8.5) is dependent on flange flexibility. The coefficients are given in Fig. 8.23. The stresses upon which Eq. (8.5) is developed agree well with results from tests of a girder with essentially rigid flanges as shown in Fig. 8.24. Note also that above the buckling load the shear stress remains constant at the buckling stress. Figure 8.25 shows that calculated values using Eq. (8.5) are usually conservative compared to test data. The data in the cases shown were from tests of riveted girders.

Figure 8.21   Failure of aluminum girder under shear loads.

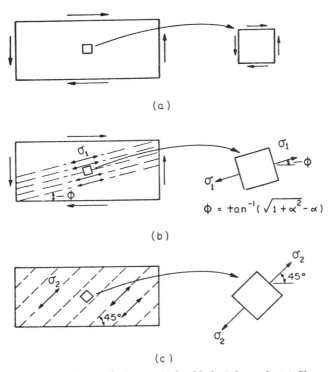

$$\phi = \tan^{-1}(\sqrt{1+\alpha^2} - \alpha)$$

Figure 8.22   Assumed stresses in buckled girder web. (*a*) Shear stresses, $\tau \leq \tau_{cr}$; (*b*) additional stresses in girder with flexible flanges; (*c*) additional stresses in girder with rigid flanges.[2]

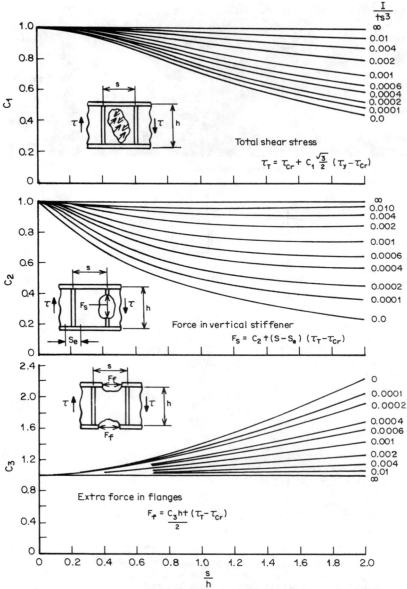

Figure 8.23 Coefficients for stresses/forces in girders. I = flexural stiffness of flange.[2]

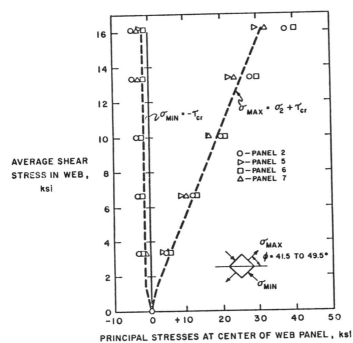

Figure 8.24 Calculated and measured stresses at the center of a panel—girder G-1.[2]

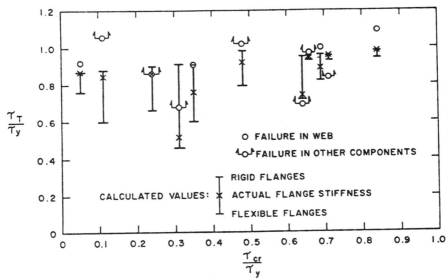

Figure 8.25 Comparison of measured and calculated shear strengths of aluminum girders.[2]

Some data are available for welded aluminum girders with relatively flexible flanges.[3] Figure 8.26 shows that Eq. (8.5) gives conservative values compared to test data. Note that the test data are for ultimate strength of the web, whereas the calculations are for yielding in the web. In these cases 10-in (254-mm) gage length yield strengths are estimated for the welded web plates, and the entire webs are assumed to have that strength.

The diagonal tension stresses in the web introduce compressive stresses in the intermediate stiffener. The force in the stiffener is as follows:

$$F_s = C_2 \, (s - s_e)t(\tau_T - \tau_{cr}) \tag{8.6}$$

where F_s = compressive force in stiffeners
$\quad s$ = spacing of transverse stiffeners
$\quad s_e = s/2$ for $s/h \le 0.3$
$\quad s_e = 0.15h$ for $s/h > 0.3$
$\quad h$ = depth of web
$\quad C_2$ = coefficient (Fig. 8.23)
$\quad t$ = thickness of web

The effective width of web sheet at the stiffener was established empirically from test data as shown in Fig. 8.27.

Additional compressive stresses are also introduced in the girder flanges as a result of the tensions developed in the web. These additional forces are

$$F_f = \frac{C_3 ht}{2} (\tau_T - \tau_{cr}) \tag{8.7}$$

where C_3 = coefficient (Fig. 8.23).

Figure 8.28 shows that the stresses calculated using Eq. (8.7) are conservative compared to data from girders with relatively stiff flanges.

8.4.2 Shear buckling of web

Practical webs have some initial out-of-flatness, and thus do not exhibit a definite buckling value. However, the theoretical buckling stress gives an approximation for the load above which diagonal tension stresses develop and deformations become large. Because most thin girder webs are stiffened, the buckling value needs to include the aspect ratio of the panels. The shear buckling of a panel may be determined by an equivalent slenderness ratio and the shear buckling

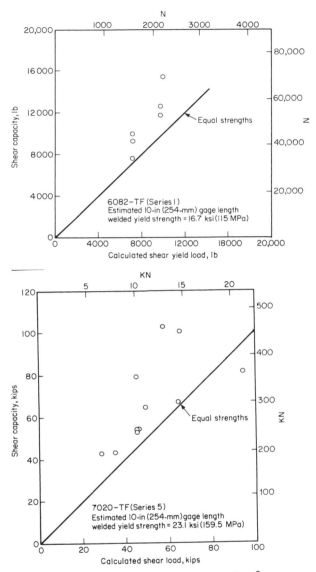

Figure 8.26   Strength of thin-web welded girders.[3]

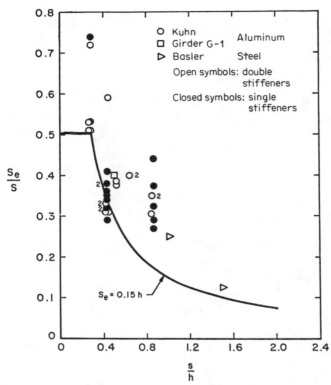

Figure 8.27 Effective width of sheet at stiffener.[2]

curves provided for long panels in shear. The equivalent slenderness ratio is as follows:[2]

$$\lambda = \frac{a}{t} \sqrt{\frac{1.6}{1 + 0.7(a/b)^2}} \qquad (8.8)$$

where λ = equivalent slenderness ratio for shear buckling
a = smaller dimension of the shear panel
b = larger dimension of the shear panel
t = thickness of web

Equation (8.8) is based on conditions at the edges that are about halfway between simple and fixed supports.

8.4.3 Intermediate stiffeners

The functions of the stiffeners on thin-web girders are to subdivide the girder webs into smaller panels, thereby increasing the shear buck-

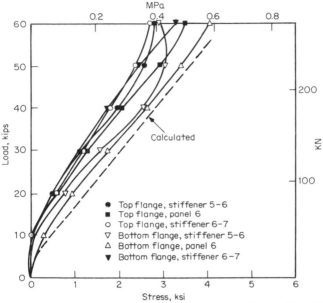

Figure 8.28 Compression stress in flange due to diagonal tension in web.[2]

ling stress, and to act as compression struts when the web stresses exceed the buckling value. The moment of inertia (about face of sheet with stiffeners on one side) recommended for these stiffeners is:[2]

For $s/h \leq 0.4$:
$$I_s = \frac{30Ds}{E}\left[1 + 0.7\left(\frac{s}{h}\right)^2\right]\left(\frac{h}{s}\right)^2 \tag{8.9}$$

For $s/h > 0.4$
$$I_s = \frac{4.8Ds}{E}\left[1 + 0.7\left(\frac{s}{h}\right)^2\right]\left(\frac{h}{s}\right)^4 \tag{8.10}$$

where $D = Et^3/12(1 - \nu^2)$
t = web thickness
ν = Poisson's ratio (1/3)
E = modulus of elasticity
s = stiffener spacing
h = clear height of web

The values as given by Eqs. (8.9) and (8.10) appear to be reasonable when compared to available design information[4] presented in Fig. 8.29. These requirements for moment of inertia are based on the local buckling stress. The form of these equations presented for design use[5] in-

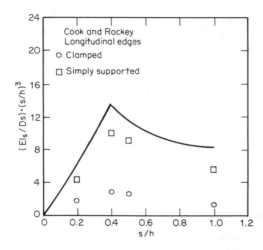

Figure 8.29 Rigidity of transverse stiffeners for girder webs.[4]

cludes the shear load so that the value of moment of inertia is dependent on load applied and not simply on the proportions of the girder web. This is probably a good design practice. To comply with this practice, the values from Eqs. (8.9) and (8.10) are multiplied by the ratio of the imposed shear load to the calculated shear load for local buckling. If large compressive stresses are calculated in the stiffeners, either from tension field action or externally applied concentrated loads, the stiffeners should be analyzed as columns or beam columns.

If the stiffeners are too thin, they will be deformed by the folds in the web and fail by forced crippling under the compression in the stiffener.[2] Figure 8.30 shows that this mode of failure may be avoided if an adequate thickness is used for the stiffener:

$$\frac{t_s}{t} = 0.65 + 0.35\sqrt{\frac{\tau}{\tau_{cr}}} \tag{8.11}$$

where t_s, t = thickness of stiffener and web plate, respectively.

8.4.4 Appearance of web

Buckle waves are generally not very apparent visually when the buckling initiates. Large folds, however, are sometimes thought to be unacceptable from the standpoint of appearance. Data are available[2] that show that appearance has been related to the slope of the buckle waves, and that the slope can be limited by restricting the amount by which the applied shear stress exceeds the theoretical buckling value. Figure 8.31 shows that to limit the slope to 4 percent or less, the shear stress must be limited to the following value:

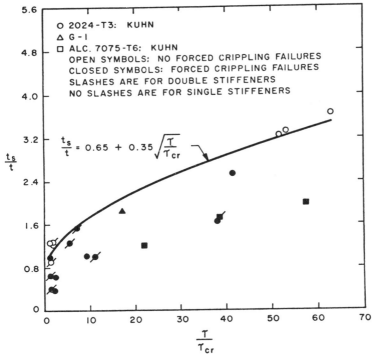

Figure 8.30 Thickness of stiffener to prohibit forced crippling failure of stiffener.[2]

$$\tau < \tau_{cr} + C_a \tag{8.12}$$

where C_a = stress added to buckling value [1.5 ksi (10.3 MPa)].

Fatigue is considered in Chap. 11. However, if cyclic loads are present in a structure with buckled elements, fatigue strength will be low compared to that for structures without buckled elements. The folds from buckling introduce high local bending stresses at joints that greatly affect fatigue. Tests will probably be needed to establish performance in this case.

8.5 Corrugated Webs and Diaphragms under Shear

Corrugated sheet has been used in trailer sidewalls, in shear diaphragms in buildings, and in various aerospace structures. The advantage of this type of construction is that the corrugations effectively act as stiffeners so that the thin sheet is very efficient in carrying

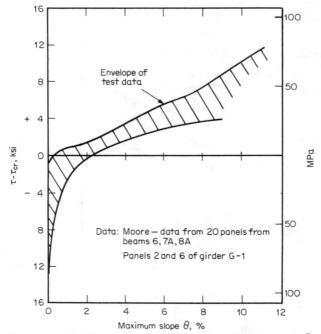

Figure 8.31   Maximum slope of buckle waves in girder webs.[2]

shear loads. The principal disadvantage arises in adequately attaching thin corrugations to edge members in a cost-effective manner.

Several modes of failure are considered:

- Overall shear buckling
- Local buckling of flats or other elements
- Failure of the attachments
- Excessive deformation

Local buckling of elements was considered in Chap. 7. This section will discuss the other three failure modes.

8.5.1   Overall shear buckling

The buckling stress of a long orthotropic plate subjected to shear has been developed elsewhere.[6] The applicable equation for a corrugated panel is the following:

$$\tau_{cr} = \frac{4K (D_x D_y^3)^{1/4}}{h^2 t} \tag{8.13}$$

where $D_x = (EI)_x/(1 - v_x v_y) \cong Ept^3/12s$ = bending stiffness in x direction

$D_y = (EI)_y/(1 - v_x v_y) \cong EAr^2$ = bending stiffness in y direction

E = modulus of elasticity

v_x, v_y = Poisson's ratios

p = pitch of corrugation

s = developed length of corrugation

A = area per unit width

r = radius of gyration per unit width

K = coefficient depending on edge fixity: 8.1 for simple supports, 15.1 for fixed edges

I = moment of inertia

Figure 8.32 gives some of the definitions of terms. Figure 8.33 shows some examples of attachment of the corrugation to flange members that seem to achieve fixed and simply supported edge conditions. For ease in designing in the elastic and inelastic regions of the material, an equivalent slenderness ratio is developed by equating Eq. (8.13) to the Euler equation. After some simplification the equivalent slenderness ratio is

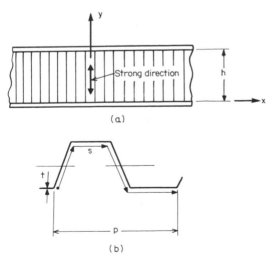

(a)

(b)

Figure 8.32 Nomenclature for corrugated webs. (a) Side view of web, (b) cross-section view.

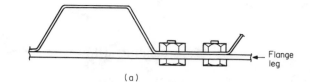

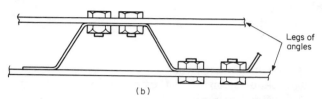

Figure 8.33 Attachment of corrugated web to flanges. (*a*) Attached at valley flat (simply supported); (*b*) attached at crown and valley flats (fixed).

$$\lambda = C \frac{h}{r} \left[\left(\frac{p}{s} \right) \left(\frac{r}{t} \right) \right]^{1/4} \tag{8.14}$$

where $C = 0.55$ for fixed edges, 0.75 for simply supported edges. One of the assumptions made in developing Eq. (8.14) is that both Poisson's ratios are zero.

Figure 8.34 shows a girder with a corrugated web that has failed by overall buckling. The corrugated sheet is a standard industrial ribbed siding. Table 8.2 gives properties of the products referred to in this section. Because the bending stiffness in the vertical direction in Fig. 8.34 is so much higher than that in the lateral direction, the buckle (damaged area) is narrow and almost vertical. Table 8.3 summarizes the girder tests made with this ribbed product. Only two of the girder webs failed in this manner. The strengths calculated using Eq. (8.14) are conservative compared to test data as shown in Fig. 8.35.

8.5.2   Failure of the attachments

Most of the failures encountered in various tests reported here occur at or near the attachments. Figure 8.36 shows a failure near the flange in which the corrugations tend to "roll over." Both crests and valleys were attached to the flanges. An equivalent slenderness ratio is suggested as follows:

$$\lambda = 2.6 \frac{a}{t} \tag{8.15}$$

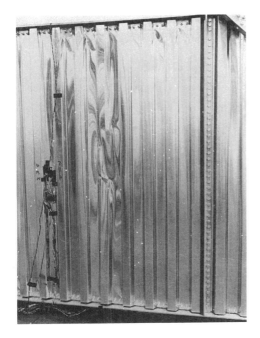

Figure 8.34 Failure of corrugated web by overall buckling.

TABLE 8.2   Properties of Industrial Products

Product	Thickness, in (mm)	Pitch of corrugation, in (mm)	Depth of section, in (mm)	Weight, lb/ft^2 (kg/m^2)	Area, in^2/ft (mm^2/cm)	Moment of inertia, in^4/ft (mm^4/cm)
Corrugated roofing and siding	0.024 (0.61)	2.67 (68)	0.88 (22.4)	0.414 (2.03)	0.352 (7.45)	0.0307 (419.2)
	0.032 (0.81)	2.67 (68)	0.88 (22.4)	0.552 (2.70)	0.469 (9.93)	0.0409 (558.5)
Ribbed siding	0.032 (0.81)	4.0 (102)	1.0 (25.4)	0.585 (2.86)	0.497 (10.52)	0.0836 (1141.6)
V-beam	0.040 (1.02)	4.88 (124)	1.75 (44.5)	0.730 (3.57)	0.621 (13.14)	0.223 (3045)

where a = width of web corrugation. The problem seems to be associated with the lack of attachment of the web of the corrugation to the flanges. The calculated values using Eq. (8.15) and the one available test point are shown in Fig. 8.37. The agreement is good but more development is needed if this type of construction is to be used.

Table 8.4 gives some results of shear tests of diaphragms that are attached to framing members with screws, more representative of the

TABLE 8.3   Test Results—Girders with Corrugated Webs

Overall depth, in (mm)	Clear depth, in (mm)	Shear stress at failure, ksi (MPa)	Type of failure	Notes
36 (914)	33 (838)	7.9 (54.5)	Bearing at flange bolts	One bolt—crest and valley
36 (914)	33 (838)	10.6 (73.1)	Crippling at end of corrugation	Two bolts—crest and valley
52 (1320)	49 (1245)	7.1 (49.0)	Bearing at flange bolts	One bolt—crest and valley
52 (1320)	49 (1245)	8.6 (59.3)	Overall buckling	Two bolts—crest and valley
52 (1320)	49 (1245)	5.6 (38.6)	Overall buckling	Two bolts—valley only
52 (1320)	49 (1245)	5.0 (34.5)	Tearing at flange bolts	Two bolts, but with hard rubber support in corrugation

4-in (102-mm) pitch industrial siding (Alc 3004-H16): thickness = 0.032 in (0.81 mm); τ_y = 19.5 ksi (134 MPa); span length of girders = 106 in (2690 mm). Load was applied at midspan.

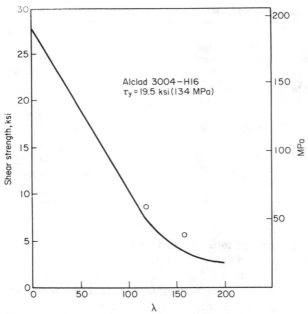

Figure 8.35   Overall buckling of corrugated webs.

Figure 8.36 Local failures of corrugations at flange attachment.

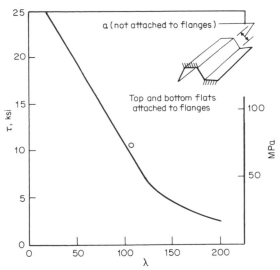

Figure 8.37 Crippling at ends of corrugations.

construction used for building products. Most of the strengths were limited by fastener-related failures. Figure 8.38 gives some information on bearing strength derived from these tests. Because of the complexity of these systems either computer simulation or tests may need to be performed for final design. Figure 8.39 shows the test setup used to obtain the above data.

TABLE 8.4 Test Results—Corrugated Panels

Product	Length, ft (m)	Purlin spacing, ft (m)	Thickness, in (mm)	Maximum shear, lb/ft (N/cm)	Initial shear stiffness, lb/in (N/mm)	Type of failure	Notes
Industrial corrugated	10 (3.05)	5 (1.52)	0.024 (0.61)	271 (39.5)	3100 (543)	Sheet buckled between fasteners.	All of the industrial corrugated panels had No. 14 self-tapping screws at 12-in (305-mm) spacing at outside edges and at 10.7-in (270-mm) spacing (in valleys) at ends and purlins.
	5 (1.52)	5 (1.52)	0.024 (0.61)	236 (34.4)	1360 (238)	Sheet buckled between fasteners and tore at end fastener.	
	10 (3.05)	5 (1.52)	0.032 (0.81)	374 (54.6)	5200 (911)	Fasteners pulled through at seam and sheet tore at end fastener.	No. 12 sheet metal screws at 12-in (305-mm) spacing were in the seams.
	5 (1.52)	5 (1.52)	0.032 (0.81)	380 (55.5)	2450 (429)	Fastener failed at end.	
V-beam	10 (3.05)	10 (3.05)	0.040 (1.02)	570 (83.2)	18,000 (3150)	Fasteners pulled through sheet at seam.	V-beam had No. 14 self-tapping screws at 12-in (305-mm) spacing on outside edges and every valley flat at ends. No. 12 sheet metal screws at 12 in (305 mm) were used at the longitudinal seams.

The width of all specimens was 8 ft (2.44 m) and three panels were used for each (two longitudinal seams).

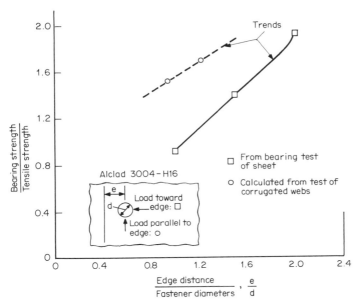

Figure 8.38   Bearing strength parallel and perpendicular to edge.

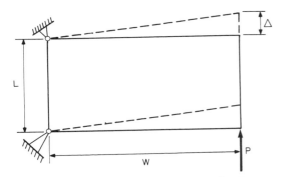

Figure 8.39   Shear deflection in corrugated web.

8.5.3 Excessive deformation

Figure 8.39 also shows the distortion of a panel subjected to shear. The equation for deflection is the following:

$$\Delta = \frac{PW}{G'L} \tag{8.16}$$

where P = shear load
   W, L = width and depth of shear panel
   G' = shear rigidity for assembly

TABLE 8.5   Constants for Deformation [Eq. (8.17)]

Product	Thickness, in (mm)	s/p	C_1, in^4 (mm^4)	C_2	Notes
Industrial ribbed	0.032 (0.81)	1.29	0 (0)	0.33	Attached crowns and valleys
Industrial ribbed	0.032 (0.81)	1.29	3.3 (1.37×10^6)	0.33	Valleys only attached
Industrial corrugated	0.024 (0.61)	1.22	102 (42.5×10^6)	16	Attached as noted, Table 8.4
	0.032 (0.81)	1.22	127 (52.9×10^6)	13	
V-beam	0.040 (1.02)	1.29	1.0 (0.42×10^6)	7.0	Attached as noted, Table 8.4

The shear rigidity needed for Eq. (8.16) is

$$G' = \frac{Gt}{s/p + [C_1/(L_c t)^2] + C_2} \tag{8.17}$$

where $G \cong \frac{3}{8}E$ = modulus of rigidity for aluminum
L_c = length in direction of corrugations
C_1, C_2 = constants (see Table 8.5)

This equation is largely empirical but it contains appropriate terms in the denominator for the various parts of the deformation. The first term corrects for the longer developed length of the sheet. The second term accounts for the rolling over of the corrugations, particularly those attached only by one flat. The last term allows for a correction for seam slip or other distortion of the fastener. Table 8.5 provides val-

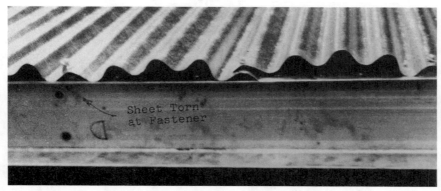

Figure 8.40   Distortions at ends of corrugated shear diaphragms.

ues for the constants for which data are available. The constants are calculated from the test data presented. Figure 8.40 shows a specimen near failure and the distortion that occurs.

8.5.4 Summary

Some general observations may be of interest. The shear stress at failure of corrugated panels failing in the local modes is essentially independent of the length parallel to the corrugations. Stiffness, however, is highly dependent on the length parallel to the corrugation.

References

1. Sharp, Maurice L., "Longitudinal Stiffeners for Compression Members," *Proceedings of the American Society of Civil Engineers*, Journal of the Structural Division, October 1966.
2. Sharp, Maurice L., and Clark, John W., "Thin Aluminum Shear Webs," *Proceedings of the American Society of Civil Engineers*, Journal of the Structural Division, April 1971.
3. Narayanan, R. (ed.), *Aluminum Structures, Design and Construction*, Elsevier Science, New York, 1987, p. 70.
4. Hartmann, E. C., and Clark, J. W., "The U. S. Code," reprint from *Proceedings, Symposium on Aluminum in Structural Engineering*, The Institution of Structural Engineers and the Aluminum Federation, London, June 11–12, 1963.
5. "Specifications for Aluminum Structures," *Aluminum Construction Manual*, 4th ed., The Aluminum Association, Washington, D.C., 1986.
6. Timoshenko, S. P., and Gere, J. M., *Theory of Elastic Stability*, McGraw-Hill, New York, 1961.

9

Cylinders and Cylindrically Curved Panels

Tubes and cylinders are efficient structural forms for carrying internal and external pressure and end compression loads. They are also often used to resist bending, especially where the loading may come from more than one direction. Some of the common uses are for light poles, framing for overhead sign trusses, piping, pressure vessels, and bulk storage tanks. This chapter covers some of the information available to the designer, but there are many details of good practice not presented here that need to be considered in each application area. Often there are user's groups that develop specific codes and standards detailing this good practice.

The information contained here pertains primarily to unstiffened construction although some effects of stiffeners on local buckling of the panels are included. The specific topics considered are as follows:

- Pipes and cylinders under internal pressure
- Pipes and cylinders under axial compression and bending
- Pipes and cylinders under shear

Buckling of cylinders under external pressure is not covered in this book. However, a discussion of this subject, along with other information on buckling of cylinders can be found in Refs. 1 and 2.

9.1 Pipes and Cylinders under Internal Pressure

The burst strength of aluminum pipe is affected by the strength of the material (longitudinal and circumferential properties) and on the ductility of the alloy.[3] *Ductility* in this case refers to the plastic deformation that can occur before rupture. The deformation increases the diameter and decreases the wall thickness of the pipe, thereby affecting the tube strength. The pressure causing failure is given by the following:[3]

$$p = \frac{K\,2t\,\sigma_t}{D - 0.8t} \qquad (9.1)$$

where p = burst pressure
K = coefficient (see Table 9.1 and Fig. 9.1)
t = thickness
D = outside diameter
σ_t = tensile strength of material in the longitudinal direction

Table 9.1 contains test data for several alloys and proportions of tubes. The length of the test specimens was about 10 times the outside diameter, and the results are the average of tests of at least three specimens. There is a tendency for the coefficient K to vary with the

TABLE 9.1 Burst Pressures in Pipe

Alloy	Outside diameter D, in (mm)	t/D	σ_y/σ_t	Burst pressure, ksi (MPa)	Calculated K
2024-0	0.625 (15.88)	0.187	0.39	12.5 (86.2)	0.93
6061-0	0.375 (9.52)	0.225	0.39	10.0 (69.0)	0.96
	0.314 (7.98)	0.154	0.34	6.10 (42.1)	0.92
	0.311 (7.90)	0.156	0.34	6.45 (44.5)	0.95
3003-F	2.384 (60.55)	0.070	0.42	2.14 (14.8)	0.87
	4.487 (113.97)	0.054	0.53	1.625 (11.2)	0.84
	8.684 (220.57)	0.038	0.52	0.981 (6.76)	0.75
3003-H14	0.376 (9.55)	0.061	0.92	3.40 (23.4)	1.05
	0.626 (15.90)	0.081	0.95	4.50 (31.0)	1.02
6061-T6	0.374 (9.50)	0.096	0.89	9.50 (65.5)	1.01
	0.998 (25.35)	0.022	0.87	2.07 (14.3)	0.95
2024-T3	1.253 (31.83)	0.028	0.71	3.75 (25.9)	0.90
2024-T4	0.250 (6.35)	0.112	0.75	16.88 (116.4)	0.93
	0.375 (9.52)	0.093	0.72	14.73 (101.6)	0.97
	0.500 (12.70)	0.098	0.72	15.20 (104.8)	0.98
	0.625 (15.88)	0.093	0.66	14.40 (99.3)	0.98
	0.750 (19.05)	0.111	0.74	17.42 (120.1)	0.96

All results are average of at least 3 tests.

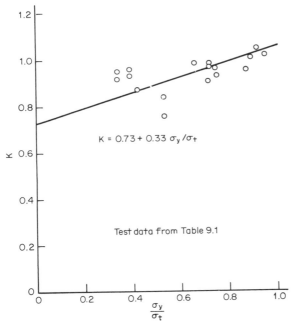

Figure 9.1   Coefficient for burst pressure in pipe.

ratio of yield to tensile strength of the material as shown in Fig. 9.1. The relationship defined is[3]

$$K = 0.73 + 0.33 \frac{\sigma_y}{\sigma_t} \tag{9.2}$$

where σ_y = yield strength of material.

Much more design information for vessels under internal or external pressure is provided by the rules of the *Boiler and Pressure Vessel Code* of the American Society of Mechanical Engineers.[1]

9.2 Cylinders under Axial Compression and Bending

The strength of thin, unstiffened cylinders loaded in axial compression is highly dependent on their imperfections. Figure 9.2 shows qualitatively the difference between the behavior of a perfect cylinder and one with imperfections. The load capacity of the cylinder with imperfections in shape is much less than that for the perfect cylinder. Also, the load capacity of the cylinder with imperfections drops off sharply after buckling. Welds also affect behavior; they cause a

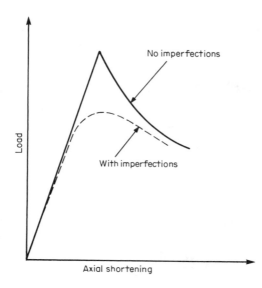

Load

No imperfections

With imperfections

Axial shortening

Figure 9.2 Load-shortening curves for unstiffened cylinders under axial compression.

reduced-strength zone in the material, and they can cause distortions that can reduce strength.

9.2.1 Cylinders under axial compression

The effects of imperfections and welding on the behavior of aluminum cylinders have been discussed in the literature.[4] The equation for elastic buckling is as follows:

$$\sigma = CE \frac{t}{R} \tag{9.3}$$

where C = coefficient that depends on degree of initial imperfections
E = modulus of elasticity
R, t = radius and thickness of cylinder

The value of the constant C is given by the following:[4]

$$C = \frac{1}{\sqrt{3(1 - \nu^2)}(1 + \sqrt{R/t}/35)^2} \tag{9.4}$$

where ν = Poisson's ratio (1/3).

Equation (9.4) is reasonable compared to test data as shown in Fig. 9.3. The coefficient approaches the theoretical value of about 0.6 for low R/t ratios and is much smaller for high R/t values. The data are for tests of relatively well-built cylinders with no damage or large imperfections. An equivalent slenderness is developed by equating Eq. (9.3)

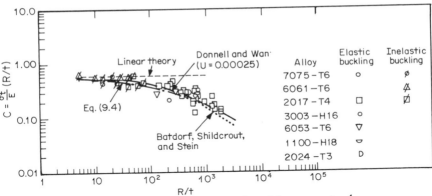

Figure 9.3   Plot of C vs. R/t for tubular members in axial compression.[4]

to the Euler equation, introducing the coefficient as given by Eq. (9.4), and solving for the slenderness ratio λ. The result is[4,5]

$$\lambda = \pi \sqrt{\frac{R}{t}} \frac{1}{\sqrt{C}} \cong 4 \sqrt{\frac{R}{t}} \left(1 + \frac{\sqrt{\frac{R}{t}}}{35} \right) \qquad (9.5)$$

For elastic buckling the equivalent slenderness is input in the Euler equation:

$$\sigma = \frac{\pi^2 E}{\lambda^2} \qquad (9.6)$$

For inelastic buckling the following equation has been proposed:[4]

$$\sigma = B_t - D_t \sqrt{\frac{R}{t}} \qquad (9.7)$$

where B_t, D_t = buckling constants for tubes in axial compression (see Table 9.2). Table 9.2 gives the coefficients for Eq. (9.7), depending on the temper of the alloy. Figure 9.4 shows that Eqs. (9.6) and (9.7) give values that are conservative compared to test data.

Welds in cylinders affect their behavior because they introduce

- A reduced-strength zone
- Imperfections
- Residual stresses

Figure 9.5 shows comparisons of test data for circumferentially welded cylinders and calculations using Eq. (9.7) and a 10-in (254-

TABLE 9.2 Buckling Constants for Tubes[4,5]

Temper	B_t / B_{tb} (ksi)	B_t / B_{tb} (MPa)	D_t / D_{tb} (ksi)	D_t / D_{tb} (MPa)	C_t
Tubes in Axial Compression:					
All alloys with tempers starting with -O, -H, -T1, -T2, -T3, -T4	$\sigma_y\left[1 + \dfrac{(\sigma_y)^{1/5}}{5.8}\right]$	$\sigma_y\left[1 + \dfrac{(\sigma_y)^{1/5}}{8.5}\right]$	$\dfrac{B_t}{3.7}\left(\dfrac{B_t}{E}\right)^{1/3}$	$\dfrac{B_t}{3.7}\left(\dfrac{B_t}{E}\right)^{1/3}$	Intersections of elastic and inelastic buckling curves are determined graphically or by trial.
All alloys with tempers starting with -T5, -T6, -T7, -T8, -T9	$\sigma_y\left[1 + \dfrac{(\sigma_y)^{1/5}}{8.7}\right]$	$\sigma_y\left[1 + \dfrac{(\sigma_y)^{1/5}}{12.8}\right]$	$\dfrac{B_t}{4.5}\left(\dfrac{B_t}{E}\right)^{1/3}$	$\dfrac{B_t}{4.5}\left(\dfrac{B_t}{E}\right)^{1/3}$	Intersections of elastic and inelastic buckling curves are determined graphically or by trial.
Tubes in Bending:					
All alloys with tempers starting with -O, -H, -T1, -T2, -T3, -T4	$1.5\sigma_y\left[1 + \dfrac{(\sigma_y)^{1/5}}{5.8}\right]$	$1.5\sigma_y\left[1 + \dfrac{(\sigma_y)^{1/5}}{8.5}\right]$	$\dfrac{B_{tb}}{2.7}\left(\dfrac{B_{tb}}{E}\right)^{1/3}$	$\dfrac{B_{tb}}{2.7}\left(\dfrac{B_{tb}}{E}\right)^{1/3}$	Intersections of elastic and inelastic curves are determined graphically or by trial.
All alloys with tempers starting with -T5, -T6, -T7, -T8, -T9	$1.5\sigma_y\left[1 + \dfrac{(\sigma_y)^{1/5}}{8.7}\right]$	$1.5\sigma_y\left[1 + \dfrac{(\sigma_y)^{1/5}}{12.8}\right]$	$\dfrac{B_{tb}}{2.7}\left(\dfrac{B_{tb}}{E}\right)^{1/3}$	$\dfrac{B_{tb}}{2.7}\left(\dfrac{B_{tb}}{E}\right)^{1/3}$	Intersections of elastic and inelastic curves are determined graphically or by trial.

TABLE 9.2 Buckling Constants for Tubes[4,5] (Continued)

	B_s		D_s		C_s
	ksi	MPa	ksi	MPa	
Tubes in Shear:					
All alloys with tempers starting with -O, -H, -T1, -T2, -T3, -T4	$\tau_y\left[1 + \dfrac{(\tau_y)^{1/3}}{6.2}\right]$	$\tau_y\left[1 + \dfrac{(\tau_y)^{1/3}}{11.8}\right]$	$\dfrac{B_s}{20}\left(\dfrac{6B_s}{E}\right)^{1/2}$	$\dfrac{B_s}{20}\left(\dfrac{6B_s}{E}\right)^{1/2}$	$\dfrac{2}{3}\dfrac{B_s}{D_s}$
All alloys with tempers starting with -T5, -T6, -T7, -T8, -T9	$\tau_y\left[1 + \dfrac{(\tau_y)^{1/3}}{9.3}\right]$	$\tau_y\left[1 + \dfrac{(\tau_y)^{1/3}}{17.7}\right]$	$\dfrac{B_s}{10}\left(\dfrac{B_s}{E}\right)^{1/2}$	$\dfrac{B_s}{10}\left(\dfrac{B_s}{E}\right)^{1/2}$	$0.41\dfrac{B_s}{D_s}$

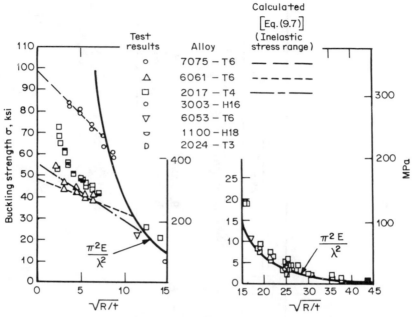

Figure 9.4 Buckling strength of round tubes in axial compression.[4]

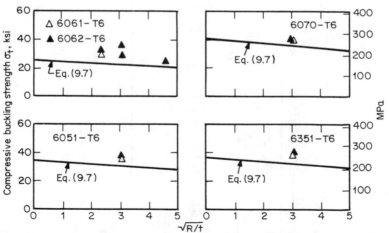

Figure 9.5 Compressive buckling strength of circumferentially groove-welded round tubes.[4]

mm) gage length yield strength. No allowance is made for either additional imperfections or residual stresses. The calculations are conservative but the data are for relatively small R/t ratios only.

Figure 9.6 shows a thin-walled cylinder with longitudinal welds after failure and Fig. 9.7 shows the failure of a cylinder of the same size, but having both longitudinal and circumferential welds. Table 9.3 gives details of the cylinders and results. An attempt was made to minimize the imperfections caused by fabrication; however, radii in some portion of each of the specimens deviated as much as 0.25 in (6.4 mm) from the nominal value. The strengths for the two cases are much different, with the cylinders with circumferential welds being much lower in strength than those with longitudinal welds. The differences were greater than those based on reduced-strength zone; thus, imperfections in shape may be the reason for the difference.

Figure 9.8 shows the results with an additional curve identified for the coefficient C, to account for the behavior of the cylinders with the circumferential welds. The equation is

$$C = \frac{1}{\sqrt{3(1 - \nu^2)}(1 + \sqrt{R/t}/11)^2} \tag{9.8}$$

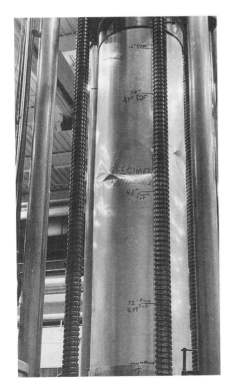

Figure 9.6 Failure of thin-walled cylinder with longitudinal welds.

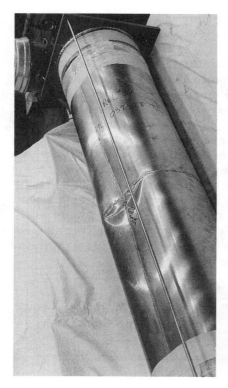

Figure 9.7   Failure of thin-walled cylinder with longitudinal and circumferential welds.

TABLE 9.3   Results of Tests of Thin-Welded 3004-H14 Cylinders under Axial Compression

Welds	Diameter, in (mm)	Thickness, in (mm)	Length, in (mm)	Maximum load, lb (N)
Two longitudinal, one circumferential	19.63 (499)	0.032 (0.81)	96 (2440)	5940 (26,400)
	20.00 (508)	0.032 (0.81)	96 (2440)	6250 (27,800)
Two longitudinal	19.76 (502)	0.032 (0.81)	90 (2290)	15,500 (68,900)
	19.74 (501)	0.032 (0.81)	96 (2440)	15,670 (69,700)

Base metal properties: tensile strength, 24.2 ksi (167 MPa); yield strength, 22.0 ksi (152 MPa). Reduced-strength area near weld has yield strength of 15.7 ksi (108 MPa); tensile strength is not measured.

The terms are defined earlier in this section. Note that Eq. (9.8) is the same as Eq. (9.4) except that the number 35 in the denominator has been replaced by 11 so that the calculated results are in agreement with the limited test data. The numbers in the denominator reflect the level of imperfections in the cylinder; the higher the number, the more perfect are the cylinders.

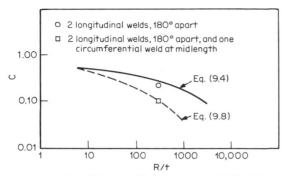

Figure 9.8 Buckling coefficients for welded thin-walled cylinders.

When this coefficient is introduced into the equation for the equivalent slenderness ratio the following results:

$$\lambda = 4\sqrt{R/t}\,(1 + \sqrt{R/t}/11) \tag{9.9}$$

The terms have been defined earlier.

Figure 9.9 shows the comparison of the test and calculated values. Based on this very limited data set, it appears that for cylinders with circumferential welds the strength at low R/t values is given by Eq. (9.7) (see Fig. 9.5), whereas Eq. (9.9) should be used for the elastic portion. Equation (9.5) seems to be reasonable for cylinders with longitudinal welds only. Note that the effects of the longitudinal welds on strength can probably be considered in the same way as described for tensile members or columns with longitudinal welds. The effect for most tanks generally will be small because the amount of unaffected

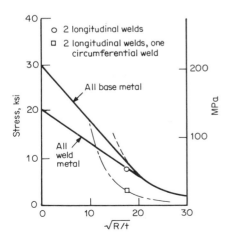

Figure 9.9 Buckling of thin-walled cylinders under axial compression.

material will be much larger than the amount of reduced-strength material.

There are two equations for the equivalent slenderness defined: Eq. (9.5), based on the performance of many typical thin-walled cylinders (no welds at large R/t ratios), and Eq. (9.9), based on limited tests of thin-walled cylinders with circumferential welds that apparently introduced larger imperfections in shape. Hopefully, these two cases represent practical limits of performance that the designer can expect in well-fabricated cylinders, although more research is needed. Until more work is done, the designer will need to select the appropriate equivalent slenderness ratio based on experience; there is no good theoretical definition available at this time to guide the designer in choosing between Eqs. (9.5) and (9.9), or some intermediate value.

Bulk storage tanks of aluminum are used for materials too corrosive to be stored in other materials, for materials that are at low temperatures, for flammable liquids and gases that need a nonsparking container, and for materials that need a container that is nonstaining. A storage tank and some of the loads that need to be considered are illustrated in Fig. 9.10. The bulk loading on the tank walls often produces circumferential tensions as well as vertical compressions. The circumferential tensions help to stabilize the tank walls and increase the compressive buckling stress.[6] Figure 9.11 shows effective R/t ratios that are inserted in the previous formulas to obtain the higher strengths. More design information for field-erected storage tanks is available.[7]

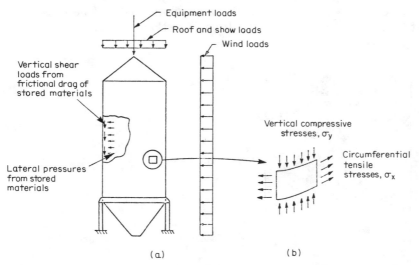

Figure 9.10 Aluminum storage tanks. (*a*) Loads on tanks, (*b*) stresses in wall.[6]

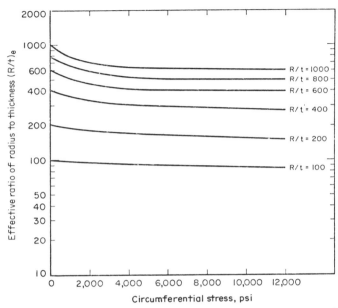

Figure 9.11   Effective values of R/t for cylinders under internal pressure.[6]

9.2.2   Cylinders under bending

The behavior of cylinders under bending is different from that for cylinders in compression in that ovaling of the cross section can occur at lower R/t ratios.[4] For cylinders with intermediate and large R/t ratios the buckling stresses are the same for compression and bending. The buckling stress for tubes and cylinders with small R/t in bending is[4]

$$\sigma = B_{tb} - D_{tb}\sqrt{\frac{R}{t}} \qquad (9.10)$$

where B_{tb}, D_{tb} = buckling constants for tubes in bending (see Table 9.2). The coefficients for Eq. (9.10) are provided in Table 9.2. The comparison between calculated and test data is given in Fig. 9.12. Figure 9.13 shows the buckling curves for one case. Three segments are needed to define behavior. The two segments to the right of that defined by Eq. (9.10) are the same as those for cylinders under axial compression.

No data are shown for thin welded cylinders in bending. The behavior is probably similar to that presented above for cylinders in axial compression.

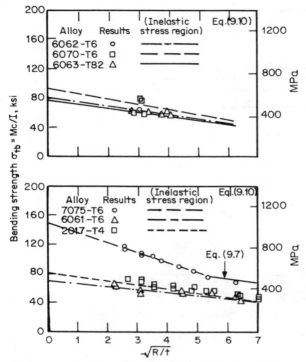

Figure 9.12   Bending strength of round tubes.[4]

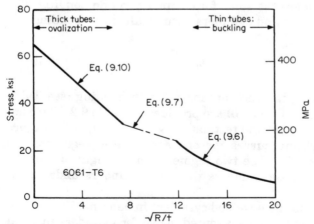

Figure 9.13   Curves for buckling-tubes in bending.

9.3   Cylinders under Torsion and Shear

The parameters relating to shear buckling of cylinders are illustrated in Fig. 9.14. Table 9.4 gives values[3] based on the equations below and on data published elsewhere, as referenced in Ref. 4. The equivalent slenderness ratio is used in equations for elastic and inelastic buckling.[4]

For elastic buckling:
$$\tau = \frac{\pi^2 E}{\lambda^2} \tag{9.11}$$

where τ = shear stress
λ = equivalent slenderness ratio (see Table 9.4)

For inelastic buckling: $\tau = B_s - D_s\lambda$ (9.12)

where B_s, D_s = coefficients given in Table 9.2. The buckling constants are the same as those given previously for buckling of flat plates. The comparison between test and calculated values is reasonable as shown in Fig. 9.15.

The equivalent slenderness ratios given in Table 9.4 can be represented by the following equation for $\lambda > \sqrt{50(R/t)}$:[3]

$$\lambda = \sqrt{40\left(\frac{R}{t}\right)^{3/2} W} \tag{9.13}$$

where $W = \begin{cases} 1.0 & \text{for } L/R \geq 9\sqrt{R/t} \\ \dfrac{1}{3}\left[\dfrac{L/R}{(R/t)^{1/2}}\right]^{1/2} & \text{for } L/R < 9\sqrt{R/t} \end{cases}$

L = length of cylinder

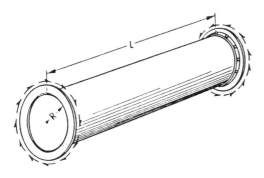

Figure 9.14 Cylinders in shear. R = mean radius of cylinder, t = thickness of shell, L = clear length between circumferential stiffeners.[3]

TABLE 9.4 Equivalent Slenderness Ratio for Tubes and Cylinders in Shear[3]

L/R	R/t =																			
	5	10	15	20	25	30	35	40	45	50	60	70	80	90	100	125	150	175	200	225
0.2	1.4	2.8	4.2	5.6	7.0	8.4	9.7	11.1	12.5	13.8	16.5	19.1	21.7	24.2	26.7	32.7	38.6	44.2	49.7	54.8
0.4	2.8	5.6	8.2	10.8	13.3	15.8	18.1	20.3	22.7	24.8	29.0	33.1	36.7	40.5	43.9	52.0	59.6	66.6	73.1	79.3
0.6	4.2	8.1	11.7	15.1	18.3	21.3	24.2	27.0	29.6	32.0	36.7	41.2	45.3	49.2	52.9	61.8	70.0	77.7	85.1	91.8
0.8	5.4	10.2	14.5	18.4	21.9	25.2	28.2	31.2	34.0	36.5	41.6	46.2	50.7	55.0	58.8	68.4	76.8	85.0	92.5	100.5
1.2	7.6	13.5	18.3	22.7	26.4	30.2	33.4	36.7	39.7	42.5	48.0	52.9	58.0	62.4	67.0	77.2	87.6	96.4	104.8	112.8
1.6	9.2	15.6	20.8	25.3	29.4	33.3	36.7	40.2	43.5	46.2	52.2	57.7	63.5	68.4	73.0	84.0	94.1	103.6	112.6	121.3
2.0	10.4	17.0	22.5	27.4	31.6	35.5	39.3	42.8	46.3	50.1	56.1	61.8	67.2	72.3	77.2	88.8	99.5	109.6	119.1	128.2
2.4	11.3	18.3	24.0	29.0	33.5	37.7	41.9	45.6	49.1	52.4	58.7	64.7	70.3	75.7	80.8	92.9	104.1	114.7	124.7	134.2
3.0	12.4	19.6	25.7	30.9	35.9	40.3	44.3	48.2	51.9	55.4	62.1	68.4	74.3	80.0	85.5	98.3	110.1	121.3	131.8	141.9
4.0	13.7	21.4	28.1	33.6	38.6	43.3	47.6	51.8	55.7	59.6	66.7	73.5	79.9	86.0	91.8	105.6	118.3	130.3	141.6	152.5
5.0	14.6	23.0	29.7	35.5	40.8	45.7	50.4	54.8	59.0	63.0	70.6	77.7	84.4	90.9	97.1	111.6	125.1	137.8	149.7	161.2
6.0	15.4	24.1	31.0	37.2	42.7	47.9	52.7	57.3	61.7	65.9	73.8	81.3	88.4	95.1	101.6	116.8	130.9	144.2	156.7	168.7
8.0	16.8	25.9	33.4	39.9	45.9	51.5	56.7	61.6	66.3	70.8	79.4	87.4	95.0	102.2	109.2	125.6	140.7	154.9	168.4	181.3
12.0	18.6	28.7	36.9	44.2	50.8	56.9	62.7	68.2	73.4	78.4	87.8	96.7	105.1	113.2	120.9	138.9	155.7	171.5	186.4	200.6
16.0	20.0	30.8	39.7	47.5	54.6	61.2	67.4	73.2	78.8	84.2	94.4	103.9	113.0	121.6	129.9	149.3	167.3	184.2	200.3	215.6
20.0	21.1	32.6	42.0	50.2	57.7	64.7	71.2	77.4	83.4	89.0	99.8	109.9	119.4	128.6	137.3	157.9	176.9	194.8	211.8	228.0
30.0	21.2	35.6	46.4	55.6	63.9	71.6	78.8	85.7	92.3	98.5	110.4	121.6	132.2	142.3	152.0	174.7	195.8	215.6	234.4	252.3
40.0	21.2	35.6	48.2	59.7	68.7	76.9	84.7	92.1	99.1	105.9	118.7	130.7	142.0	152.9	163.3	187.7	210.4	231.7	251.8	271.1
50.0	21.2	35.6	48.2	59.8	70.7	81.1	89.6	97.4	104.8	112.0	125.5	138.2	150.2	161.7	172.7	198.5	222.5	245.0	266.3	286.6
60.0	21.2	35.6	48.2	59.8	70.7	81.1	91.1	100.6	109.7	117.2	131.3	144.6	157.2	169.2	180.7	207.8	232.8	256.4	278.7	300.0

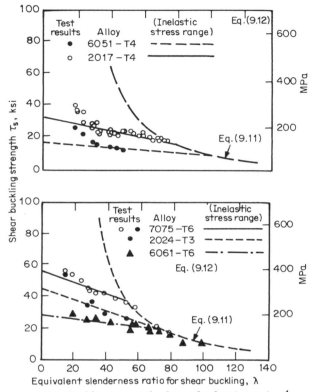

Figure 9.15 Buckling strength of round tubes in torsion.[4]

R and t are as defined previously. This slenderness ratio is used in Eqs. (9.11) and (9.12).

The addition of longitudinal stiffeners to the cylinder can increase the shear buckling stress.[3] The slenderness ratio for longitudinally stiffened cylinders is

$$\lambda_s = \frac{\lambda}{\sqrt{1 + \frac{1}{2}(t/s)^2 \, \lambda^2}} \tag{9.14}$$

where λ_s = equivalent slenderness ratio for tube and cylinder in shear with longitudinal stiffeners

λ = equivalent slenderness ratio for tube and cylinder in shear (no longitudinal stiffeners) [Table 9.4 or Eq. (9.13)]

s = spacing of longitudinal stiffeners

t = thickness of plate

Equation (9.14) is inserted in Eqs. (9.11) and (9.12).

9.4 General

The cases discussed above concern buckling of the cylinder. For unstiffened cylinders the buckling load is the largest practical load that can be applied. Stiffened structures, however, may have strengths higher than the buckling load depending on their proportions. The mechanisms that allow these higher loads are similar to those described previously for postbuckling strength of flat plates.

References

1. "Section VIII Rules for Construction of Pressure Vessels," Division 1, *ASME Boiler and Pressure Vessel Code,* American Society of Mechanical Engineers, New York, 1989.
2. Galambos, Theodore V. (ed.), *Guide to Stability Design Criteria for Metal Structures,* Wiley, New York, 1988.
3. *Alcoa Structural Handbook,* Aluminum Company of America, Pittsburgh, Pennsylvania, 1960.
4. Clark, John W., and Rolf, Richard L., "Design of Aluminum Tubular Members," *Proceedings of the American Society of Civil Engineers,* Journal of the Structural Division, December 1964.
5. "Specifications for Aluminum Structures," *Aluminum Construction Manual,* Sec. 1, The Aluminum Association, Washington, D.C., 1986.
6. "How Rugged Should Your Bulk Bin Be?" *Modern Materials Handling,* May 1970.
7. *Welded Aluminum-Alloy Storage Tanks,* ASME/ANSI B96.1, 1986.

Joint Behavior

Most of the joining methods utilized for aluminum members are the same as those employed for other materials. This chapter discusses some of the common methods available: welding, mechanical fastening, and adhesive bonding. The information presented is limited to structural behavior and design. The processes used to make the joints are always important: if they are not properly executed, the joints will not perform well. References on process are included here when possible. Because fatigue is so important in the design of many structures and the joints are primary locations for failures, an entire chapter will be devoted to that subject; thus, fatigue is not covered here.

10.1 Welded Joints

The common practices employed for aluminum weldments are gas metal arc welding (GMAW), also given the acronym MIG, and gas tungsten arc welding (GTAW), also known as TIG.[1,2] The design data presented apply to either of these methods. The filler wire recommended for these processes depends on the base alloy, as shown in Table 10.1. The choice of filler alloy, when more than one is possible, is based both on process and performance considerations. In addition, the behavior of resistance spot welds is discussed; this method is used for attaching thin sheets together, particularly in the automotive industry. Also, some data are included related to the postweld treatment of the joint to improve properties.

10.1.1 Groove-weld behavior

Chapter 5 presented strengths of various types of members with groove (or butt) welds. Table 10.2 presents the data given in that chapter for welds stressed normal to their axes. Also included in Chap.

TABLE 10.1 Aluminum Filler Alloys Recommended for General-Purpose Welding (Base Metal to Base Metal)[1]

Base metal	Base metal										
	356.0, A356.0, A444.0	6005, 6061, 6063, 6351	5456	5454	5154, 5254	5086	5083	5052, 5652	5005, 5050	3004, Alc. 3004	1100, 3003, Alc. 3003
1100, 3003, Alclad 3003	ER4043[a,b]	ER4043[a]	ER5356[c]	ER4043[a,c]	ER4043[a,d]	ER5356[c]	ER5356[c]	ER4043[a,d]	ER4043[d]	ER4043[d]	ER1100[d]
3004, Alclad 3004	ER4043[a]	ER4043[e]	ER5356[d]	ER5654[e]	ER5654[e]	ER5356[d]	ER5356[d]	ER4043[a,d]	ER4043[d]	ER4043[d]	—
5005, 5050	ER4043[a]	ER4043[e]	ER5356[d]	ER5654[e]	ER5654[e]	ER5356[d]	ER5356[d]	ER4043[a,d]	ER4043[d,f]	—	—
5052	ER4043[a,e]	ER5356[c,e]	ER5356[e]	ER5654[e]	ER5654[e]	ER5356[d]	ER5356[d]	ER5654[c,e]	—	—	—
5083	ER5356[a,c,d]	ER5356[d]	ER5183[d]	ER5356[d]	ER5356[d]	ER5356[d]	ER5183[d]	—	—	—	—
5086	ER5356[a,e]	ER5356[d]	ER5356[d]	ER5356[e]	ER5356[e]	ER5356[d]	—	—	—	—	—
5154	ER4043[a,e]	ER5356[c,e]	ER5356[e]	ER5654[e]	ER5654[e]	—	—	—	—	—	—
5454	ER4043[a,e]	ER5356[c,e]	ER5356[e]	ER5554[c,d]	—	—	—	—	—	—	—
5456	ER5356[a,e]	ER5356[d]	ER5556[d]	—							
6005, 6061, 6063, 6351	ER4043[a,e]	ER4043[a,e]	—								
356.0, A356.0, A444.0	ER4043[a,f]	—									

[a] ER4047 may be used for some applications.

[b] ER4145 may be used for some applications.

[c] ER4043 may be used for some applications.

[d] ER5183, ER5356, or ER5556 may be used.

[e] ER5183, ER5356, ER5554, and ER5654 may be used. In some cases, they provide (1) improved color match after anodizing treatment, (2) highest weld ductility, and (3) higher weld strength. ER5554 is suitable for elevated-temperature service.

[f] Filler metal with the same analysis as the base metal is sometimes used.

TABLE 10.2 Minimum Mechanical Properties for Welded Aluminum Alloys[3]

Alloy and temper	Product	Thickness range, in (mm)	Tension		Compression	Shear		Bearing	
			Ultimate,* ksi (MPa)	Yield,† ksi (MPa)	Yield,† ksi (MPa)	Ultimate, ksi (MPa)	Yield, ksi (MPa)	Ultimate, ksi (MPa)	Yield, ksi (MPa)
1100-H12, H14	All	All	10 (70)	4.5 (30)	4.5 (30)	8 (55)	2.5 (15)	23 (160)	8 (55)
Alclad 3004-H16	All	All	19 (130)	11 (75)	11 (75)	13 (90)	6.5 (45)	44 (305)	19 (130)
5052-H34	All	All	22 (150)	13 (90)	13 (90)	16 (110)	7.5 (50)	50 (345)	19 (130)
5083-H321	Sheet and plate	0.188–1.500 (4.8–38)	36 (250)	24 (165)	24 (165)	24 (165)	14 (95)	80 (550)	36 (250)
5086-H116	Sheet and plate	All	32 (220)	19 (130)	19 (130)	21 (145)	11 (75)	70 (480)	28 (195)
5454-H34	Sheet and plate	All	28 (195)	16 (110)	16 (110)	19 (130)	9.5 (65)	62 (425)	24 (165)
5456-H116	Sheet and plate	0.188–1.500 (4.8–38)	38 (260)	26 (180)	24 (165)	25 (170)	15 (105)	84 (580)	38 (260)
6061-T6, T62‡	All	All‡	22 (150)	20 (140)	20 (140)	15 (105)	12 (85)	50 (345)	30 (210)
6061-T6, T62 (welded with 4043, 5554, 5654)	All	Over 0.375 (9.5)	22 (150)	15 (105)	15 (105)	15 (105)	9 (60)	50 (345)	30 (210)
6063-T5	All	All	15 (105)	11 (75)	11 (75)	11 (75)	6.5 (45)	34 (235)	22 (150)

*90% of ASME weld qualification tests.
†Corresponding to 0.2% offset in a 10-in (254-mm) gage length across a butt weld.
‡See values below for thicknesses over 0.375 in (9.5 mm) when filler alloys are 4043, 5554, or 5654.

5 were methods for evaluating the strength of welded members when they are stressed parallel to the axis of the weld. Following are data on more practical members with the emphasis on the joints.

Figure 10.1 shows tensile and yield strengths of various types of members of 6061-T6 with groove welds.[4] There are some differences in strength, but the values as given in Table 10.2 are conservative compared to the test data. For example, specimens with a filler alloy of 4043 had lower strength than those made with 5356 or 5556. Values for yield strength in Table 10.2 are lower for groove welds of 4043 in material over ⅜ in (9.5 mm) compared to those for the other recommended fillers. In a few cases a 10-in (254-mm) gage length yield strength is estimated by multiplying the value for a 2-in (50.8-mm) gage length yield strength by ⅓.

Similar data are provided for groove welds in the tensile flange of beams in Fig. 10.2.[5] To obtain the calculated values in Fig. 10.2, the strengths from Table 10.2 are multiplied by shape factors, estimated by the procedure defined in Chap. 6. The calculated results are conservative compared to test data.

The data provided above and in Chap. 5 cover only those cases in which the applied stress is parallel to or perpendicular to the axis of the weld. Figure 10.3 shows failures of butt welds oriented 90, 45, and 25° to the axis of the specimen and the loading direction. The test data are provided in Table 10.3. There is not much difference in the

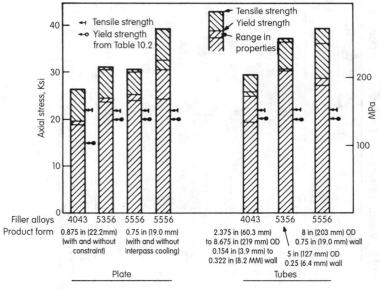

Figure 10.1 Groove welds, 6061-T6.[4]

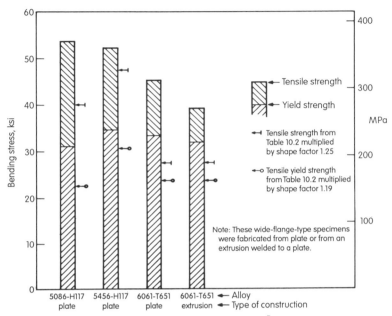

Figure 10.2   Groove welds in tensile flange of beams.[5]

Figure 10.3   Failure of inclined butt welds.

TABLE 10.3   Results of Tests of Inclined Butt Welds*

Inclination of weld to longitudinal axis, °	Tensile strength, ksi (MPa)	Yield strength,† ksi (MPa)	Elongation, % in 10 in (254 mm)
90	34.4 (237)	27.6 (190)	2.0
45	33.2 (229)	25.4 (175)	3.5
25	36.2 (250)	25.9 (179)	6.9

Base metal properties: tensile strength = 45.6 ksi (314 MPa), yield strength = 42.9 ksi (296 MPa), elongation in 2 in (50.8 mm) = 15%.
*Specimens 4.28 in (109 mm) wide by 0.375 in (9.5 mm) thick, 5356 filler.
†Stress at 0.2% offset in 10-in (254-mm) gage length.

strength of the parts. Also, note that the failure analysis needs to take into account the various fracture patterns shown. A gross overestimation of strength would result if the designer simply calculated the strength of a section cut across the member using the formulas for members with longitudinal welds. The inclined welds did increase the elongation of the joint significantly. Thus, inclined welds may be useful to increase deformation capacity and thus energy absorption.

"Structural toughness" is important for ship structures, automotive frames, and other similar applications. Good toughness is achieved if the structure is capable of undergoing large inelastic deformations, and, thus, absorbs large amounts of energy. Often the joint limits the amount of energy that the structure can absorb. Figure 10.4 shows the relative amount of energy absorbed to failure by beams with and with-

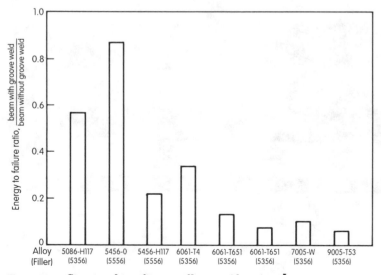

Figure 10.4   Structural toughness—alloy considerations.[5]

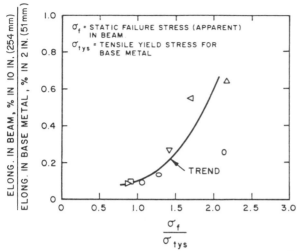

Figure 10.5 Deformation in beams with transverse welds.[5]

out transverse groove welds. The relative energies depend on alloy and temper. Figure 10.5 shows that the strength of the joint must be well above the yield strength of the base metal to get good energy absorption. If not, all of the inelastic deformation (and most of the energy absorbed) occurs in the local region of the joint, while the base metal is only elastically deformed. Proper alloy selection and design are needed to achieve good energy absorption if required. Many applications do not need tough structures. The derrick for the Alcoa *Seaprobe* mentioned in Chap. 2 is an example where strength is controlling. Breakaway devices for the bases of light poles must have low energy absorption.

10.1.2 Fillet-weld behavior

Figure 10.6 shows specimens used to obtain basic information on fillet-weld strength.[6] Figures 10.7, 10.8, and 10.9 reproduce the data for several common filler alloys.[6] Strengths for transverse fillet welds (loading perpendicular to the axis of the weld) generally are significantly higher than those for longitudinal fillet welds apparently because of the different type of stress in the fillets in the specimens employed. There are differences in strength depending on filler alloy. In these tests there was no differentiation depending on base alloy. Table 10.4 gives minimum proposed values for design.[6]

Figure 10.10 defines failure modes that can occur in fillet-weld connections. The upper sketch shows the normal failure through the fillet weld. The lower sketch shows failure possibilities at the interface be-

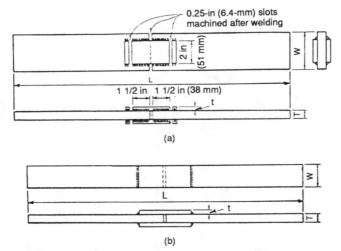

Figure 10.6 Fillet-weld specimens. (*a*) Longitudinal shear specimen, (*b*) transverse shear specimen.[6]

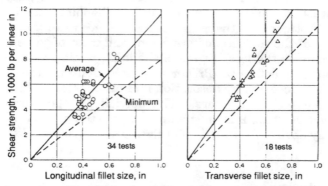

Figure 10.7 Shear strengths of fillet welds made with 4043 filler metal naturally aged (2–3 months).[6]

tween the fillet and base metal; these failures are most likely to occur if a strong filler alloy is used to weld a lower-strength alloy. The designer should check all the possible modes of failure. The interface shear strength can be assumed to be equal to the ultimate shear strength as given in Table 10.2. Figure 10.11 shows data for tests in which the base metal is kept constant and the filler alloy is varied to obtain interface failures. Most of the failure loads in this case are limited by interface strength. Figure 10.12 shows evidence of interface and fillet failures in specimens in which filler alloy was fixed and the

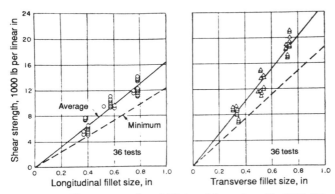

Figure 10.8 Shear strengths of fillet welds made with 5356 filler metal.[6]

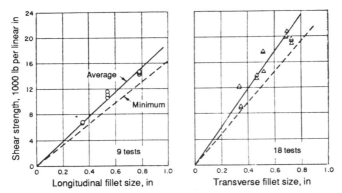

Figure 10.9 Shear strengths of fillet welds made with 5556 filler metal.[6]

TABLE 10.4   Minimum Shear Strengths for Fillet Welds[6]

Filler alloy	Shear strength, ksi (MPa)	
	Longitudinal	Transverse
4043*	11.5 (79)	15 (103)
5356	17 (117)	26 (179)
5554	17 (117)	23 (159)
5556	20 (138)	30 (207)

Shear strength values are obtained by dividing the load at failure by the net section of the fillets.
*Naturally aged (2–3 months).

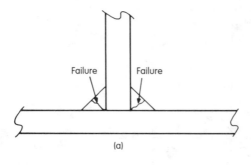

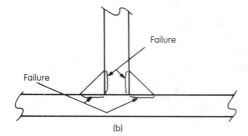

Figure 10.10 Failure modes for fillet welds. (*a*) Failure through net section, (*b*) failure through base metal at interface.

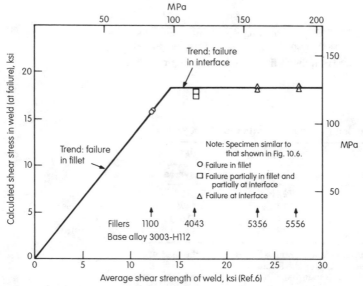

Figure 10.11 Failure of specimens with longitudinal fillet welds.

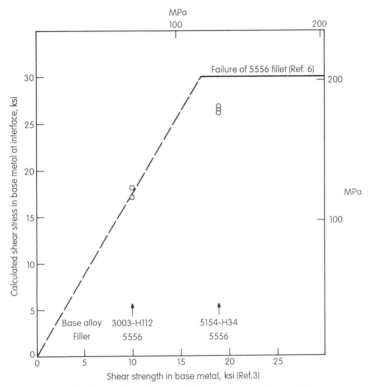

Figure 10.12 Failure of specimens with transverse fillet welds.

base alloy changed. The shear strengths of the base alloys were obtained from minimum expected as-welded properties.[3]

Results of tests of specimens with transverse fillet welds on various tubular members and plate specimens, as shown in Fig. 10.13, reveal that the strengths for the tubular joints are much less than those given in Table 10.4. The failures were reported to be in the welds. The calculated stresses at the interface also were high. Either the interface stresses were a factor or the stresses in the joint were much different from those in the plate specimens. Stresses are expected to be different because the welds in the tubular joints are unsymmetrical (one side of wall only) and thus have bending stresses, whereas the plate specimens have symmetrical welds (Fig. 10.6) and no bending stresses. The designer needs to utilize the lower values for transverse welds in tubular joints (about one-half those for transverse welds in plate specimens).

Figure 10.14 shows that strengths of longitudinal fillet welds for gusset plates attached to tubes[4] are lower than those for plate

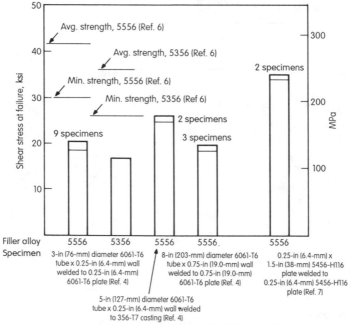

Figure 10.13   Strength of transversely loaded fillet welds.

specimens,[6] but those for cruciforms[7] are about the same. The reasons for the differences—stress condition or process differences, for example—are not known at present. The designer needs to use conservative values for practical structures.

10.1.3   Resistance spot-weld behavior

Spot welds are often used to join thin sheets. Guides on process, size, spacing, and strength are already available.[8] Figure 10.15 summarizes some of the requirements and gives some test data for several alloys and tempers (upper part). The strength guidelines are conservative. The test results are from various types of joints containing from one to four spots.

10.1.4   Effects of postweld treatments

The properties of as-welded joints may be changed by thermal processing after welding. For example, manufacturers are allowed to increase the design strength of welded joints to 85 percent of base metal properties for 6005-T5 and 6063-T6 lighting-pole assemblies by precipitation heat-treating (artificial aging) after welding.[3] Welding is done with 4043 filler with the poles of certain thicknesses and in the proper

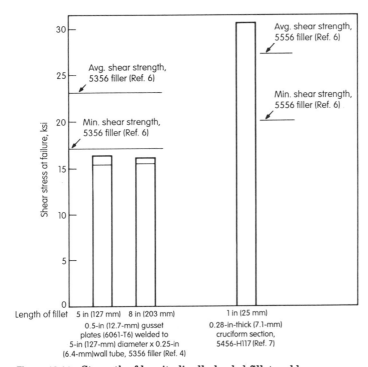

Figure 10.14   Strength of longitudinally loaded fillet welds.

temper [less than 0.25 in (6.4 mm) with 6005-T1 and less than 0.375 in (9.5 mm) with 6063-T4].

Figure 10.16 shows data for beam behavior for several conditions of as-welded and postweld treatment.[9] The base condition is for a beam with longitudinal welds in the as-welded condition. The deflection to failure and strength to failure for this case are considered equal to 1.0. The strength was slightly increased, but the deformation to failure was greatly reduced by re-heat-treating and aging after welding. Beams with a transverse weld had much lower strength and deflection to failure than the base case, as expected. Postweld heat treatment and aging improved strength but reduced deflection to failure. In general, proper postweld treatment should improve strength but probably reduces structural toughness.

10.2   Mechanically Fastened Joints

Fasteners used to create the joints in this category are aluminum bolts, rivets, mechanical clinches, and snap-fit attachments. Also, gal-

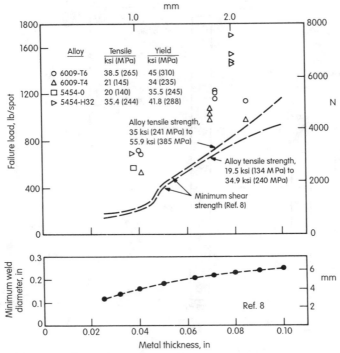

Figure 10.15   Strength of spot welds.

vanized steel bolts and 300 series stainless steel bolts may be used in place of aluminum bolts. The selection of the fastener is based on structural requirements and, in corrosive environments, on the prevention of galvanic corrosion. Mechanical fasteners probably are the most reliable and trouble-free method of joining aluminum. Visual inspection will identify most installation problems. References to proper installation practices will be provided where available.

10.2.1   Types and selection of fasteners

Riveting is extensively used as a method for joining aluminum parts; aircraft structures, tractor trailers, and railway cars are a few examples. Table 10.5 provides some guidelines for the selection of rivet alloy.[10] One of the criteria of selection is to prevent galvanic corrosion in very corrosive environments, such as salt water. The rivet should have about the same galvanic relationship or be slightly cathodic to the surrounding material.

Some of the types of rivets available are shown in Fig. 10.17. Rivets are normally used to transmit shear loads only. The shear capacity is

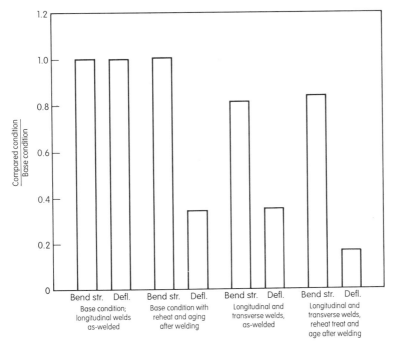

Figure 10.16   Effect of postweld treatment on welded beam behavior: 6061-T6 beams, 4043 filler.[9]

TABLE 10.5   Some Satisfactory Combinations of Aluminum Structural Alloys and Aluminum Fastener Alloys[10]

Structural alloys	Rivet alloys	Threaded-fastener alloys
1XXX series	1100	2024
2XXX series	2017, 2024, 2117, 6061, 7050	2024, 7050, 7075
3XXX series	6053, 6061	2024, 7050, 7075
5XXX series	6053, 6061	6061, 7050, 7075
6XXX series	6053, 6061	2024, 6061, 7050, 7075
7XXX series	2017, 2024, 2117, 6061, 7050	2024, 6061, 7050, 7075

SOURCE: Reprinted with permission from SAE Paper No. 800455, © 1980 Society of Automotive Engineers, Inc. (Ref. 10).

dependent on the area of the rivet and the shear strength of the fastener materials as driven. Average shear strengths after driving are given in Table 10.6 for some of the common rivet alloys. Good practices for the installation of the rivets are available.[3,10,11]

Aluminum bolts, screws, and other threaded fasteners are also commonly used. Table 10.7 provides minimum tensile and shear strengths

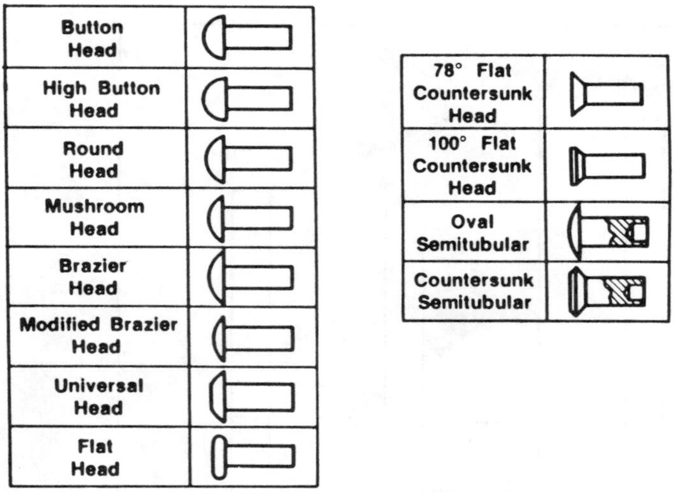

Button Head	
High Button Head	
Round Head	
Mushroom Head	
Brazier Head	
Modified Brazier Head	
Universal Head	
Flat Head	

78° Flat Countersunk Head	
100° Flat Countersunk Head	
Oval Semitubular	
Countersunk Semitubular	

Figure 10.17   Some common head styles for aluminum rivets. (*Reprinted with permission from SAE Paper No. 800455, © 1980 Society of Automotive Engineers, Inc.*)

TABLE 10.6   Average Shear Strength of Driven Rivets[10]

Alloy and temper before driving	Driving procedure	Temper after driving	Shear strength, ksi (MPa)
1100*	Cold, as-received	F	11 (76)
2017-T4	Cold, as-received	T3	39 (269)
2017-T4	Cold, immediately after quenching or refrigerated	T31	34† (234)
2024-T4	Cold, immediately after quenching or refrigerated	T31	41† (283)
2117-T4	Cold, as-received	T3	33 (228)
6053-T61	Cold, as-received	T3	23 (159)
6061-T6	Cold, as-received	T6	30 (207)
7050-T73	Cold, as-received	T73	45 (310)

*Rivets are manufactured from 1100-H14 rivet wire or rod.
†Immediately after driving, the shear strength of these rivets is about 75% of the values shown. On standing at ordinary temperatures, they age-harden to develop their full shear strength, this action being completed in about four days.
SOURCE: Reprinted with permission from SAE Paper No. 800455, © 1980, Society of Automotive Engineers, Inc. (Ref. 10).

of some of the alloys used. Table 10.8 gives nominal sizes, types of thread, and the major (gross) and minor (net at threads) diameters. Figure 10.18 shows that the strength of the fastener (tension in this case) is conservatively calculated from the strength of the alloy and the appropriate net area of the fastener.

The torque to fail the fastener depends on the assembly being fas-

TABLE 10.7 Minimum Tensile and Shear Strength of Aluminum Alloys Used for Threaded Fasteners[10]

Alloy and temper	Tensile strength, ksi (MPa)	Shear strength, ksi (MPa)
2024-T4	62 (427)	37 (255)
6061-T6	42 (290)	25 (172)
6262-T9*	50 (345)	29 (200)
7050-T73	70† (483)	41† (283)
7075-T6	77 (531)	42 (290)
7075-T73	68 (469)	41 (283)

*Used for nuts only.
†Tentative values.
SOURCE: Reprinted with permission from SAE Paper No. 800455, © 1980, Society of Automotive Engineers, Inc. (Ref. 10).

TABLE 10.8 Nominal Dimensions of Threaded Fasteners[10]

Nominal size	Threads per inch*	Diameter	
		Major, in (mm)	Minor, in (mm)
4	40 UNC†	0.112 (2.84)	0.0805 (2.04)
	48 UNF‡	0.112 (2.84)	0.0857 (2.18)
5	40 UNC	0.125 (3.18)	0.0935 (2.37)
	44 UNF	0.125 (3.18)	0.0964 (2.45)
6	32 UNC	0.138 (3.51)	0.0989 (2.51)
	40 UNF	0.138 (3.51)	0.1065 (2.71)
8	32 UNC	0.164 (4.17)	0.1284 (3.26)
	36 UNF	0.164 (4.17)	0.1291 (3.28)
10	24 UNC	0.190 (4.82)	0.1379 (3.50)
	32 UNF	0.190 (4.82)	0.1508 (3.83)
12	24 UNC	0.216 (5.49)	0.1639 (4.16)
	28 UNF	0.216 (5.49)	0.1712 (4.35)
1/4	20 UNC	0.250 (6.35)	0.1876 (4.76)
	28 UNF	0.250 (6.35)	0.2052 (5.21)
5/16	18 UNC	0.3125 (7.94)	0.2431 (6.17)
	24 UNF	0.3125 (7.94)	0.2603 (6.61)
3/8	16 UNC	0.375 (9.52)	0.2970 (7.54)
	24 UNF	0.375 (9.52)	0.3228 (8.20)
1/2	13 UNC	0.50 (12.7)	0.4056 (10.30)
5/8	11 UNC	0.625 (15.9)	0.5135 (13.04)
3/4	10 UNC	0.750 (19.0)	0.6273 (15.93)
7/8	9 UNC	0.875 (22.2)	0.7387 (18.76)
1	8 UNC	1.000 (25.4)	0.8466 (21.50)

*Class 2A, external threads.
†UNC = Unified National Coarse Thread Series.
‡UNF = Unified National Fine Thread Series.

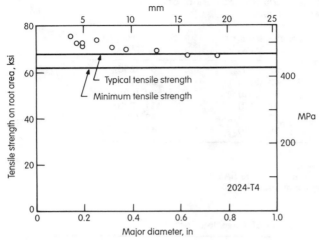

Figure 10.18 Tensile strength of threaded fasteners.[10]

tened and the lubrication of the threads and surface between nut and washer. A suggested technique for establishing an installation torque is to break at least three bolts in torsion in an assembly representative of the final structure. The installation torque should be 80 percent of the least torque obtained from the test.

10.2.2 Behavior of mechanical joints in structures

One of the common types of failure for thin roofing and siding products is "pull-through": the fastener pulls through the sheet. Figure 10.19 shows data for pull-through strengths of several washer types used with a No. 14 stainless steel fastener. The material properties for the formed sheet products are given in Table 10.9.

The pull-through strength is given roughly by the following:

$$P = Ct\sigma_t(D - d) \tag{10.1}$$

where P = pull-through strength
   C = coefficient (1.0 for valley fastening and 0.7 for crown installation)
   σ_t = tensile strength of the sheet
   t = thickness of sheet
   D = diameter of washer
   d = diameter of fastener

Note that Eq. (10.1) is approximate and has not been confirmed except for the specific cases shown.

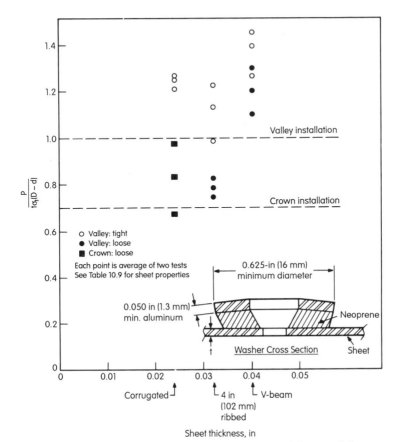

Figure 10.19 Pull-through strengths using No. 14 stainless steel fastener with washer.

TABLE 10.9   Material for Pull-Through Strength Tests*

Product	Thickness, in (mm)	Tensile strength, ksi (MPa)	Yield strength, ksi (MPa)	Elongation in 2 in (50.8 mm), %
Corrugated industrial sheet	0.024 (0.61)	35.65 (246)	33.15 (229)	3.2
4-in (102-mm) pitch-ribbed industrial siding sheet	0.032 (0.81)	33.75 (233)	30.5 (210)	2.2
V-beam roofing and siding	0.040 (1.02)	35.30 (243)	31.4 (217)	4.0

*Alloy is 3004-H16.

There are a number of potential failure modes that can occur in a bolted connection subjected to shear loading. They include

- Shear failure of the fastener
- Bearing failure of the sheet or plate
- Tensile failure at the net section

The properties of the fasteners and the sheet and plate have been provided previously in this book. The bearing properties are based on tests in which the distance from the center of the fastener to the edge of the material (edge distance) toward which the pressure is directed is twice the diameter of the fastener. If the distance is less, the bearing values are reduced by the following ratio: actual edge distance/twice the fastener diameter. If the bearing strength is divided by the tensile strength, the relationship may be expressed as follows:

$$\sigma_b/\sigma_t = (e/2d)(\sigma_b/\sigma_t)_{max} \qquad (10.2)$$

where σ_b/σ_t = ratio of bearing strength to tensile strength
      e = distance from center of hole to edge of sheet
      d = diameter of hole
      $(\sigma_b/\sigma_t)_{max}$ = ratio when $e \geq 2d$

Figure 10.20 shows good agreement of data introduced in Chap. 8 and Eq. (10.2).

When the pressure applied by the fastener is parallel to the edge, the reduction can be less, as shown in Fig. 10.21 and the following equation:

$$\sigma_b/\sigma_t = 0.9 + 0.6e/d \qquad (10.3)$$

The terms are defined above. Equation (10.3) is empirical and based on limited data.

An additional failure mode can occur in gusset plates and angles attached by one leg, as illustrated in the upper sketch of Fig. 10.21. The lower right-hand corner can tear off. The strength is estimated by summing the strengths in shear and tension. The lower sketch shows a reduced section needed to estimate tensile strength because the loading is not uniform (note that some design guidelines[3] suggest using only one-third of the width of the unconnected leg). All the failure modes are compared with test data for 6061-T6 angles attached by one leg in Fig. 10.22. The upper lines are typical and the lower lines are minimum-strength values. Some details of the test setup are provided in Chap. 5.

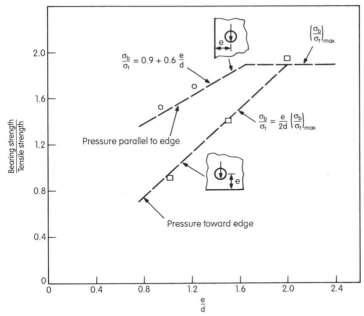

Figure 10.20 Bearing failures in joints.

Aluminum fasteners are used in many applications and perform well. In heavy construction, however, higher-strength fasteners may be needed in order to pull parts together. An example is the attachment of the deck section on the Smithfield Street Bridge in Pittsburgh to the floor beams. The deck sections were made to match the contour of the existing floor beams (see Fig. 10.23), but there was not a perfect fit, and the 5/8-in (15.9-mm) aluminum bolts were not strong enough to eliminate the small mismatch at the attachment. The lack of tightness was not obvious by visual inspection, but it allowed the bolts to be cyclically unloaded by truck loads, resulting in fatigue failures of some of the aluminum bolts after a few years of operation. Galvanized A356 steel bolts were used as the replacement, and these bolts eliminated the mismatch, thus solving the problem.

10.2.3 Mechanical clinches

For thin sheet, there are commercial types of clinches that locally deform and/or shear the metal sufficiently to attach sheet together. Figure 10.24 gives data for one of the common types. These fasteners have been used to attach parts of automotive seat structures together.

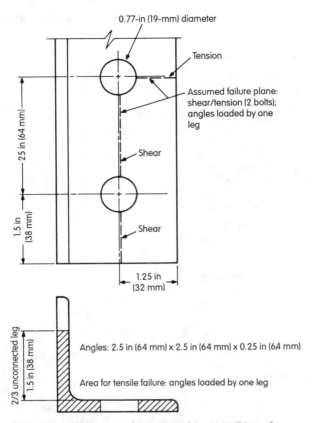

Figure 10.21 Failure modes assumed in 6061-T6 angle.

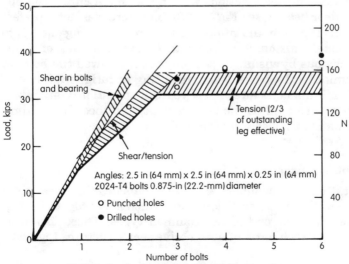

Figure 10.22 Failure in single angles attached by one leg.

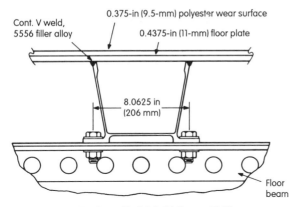

Cont. V weld,
5556 filler alloy

0.375-in (9.5-mm) polyester wear surface

0.4375-in (11-mm) floor plate

8.0625 in
(206 mm)

Floor
beam

Figure 10.23   Deck on Smithfield Street Bridge.

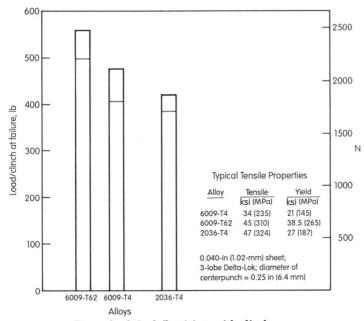

Figure 10.24   Strength of single-lap joints with clinches.

Typical Tensile Properties

Alloy	Tensile ksi (MPa)	Yield ksi (MPa)
6009-T4	34 (235)	21 (145)
6009-T62	45 (310)	38.5 (265)
2036-T4	47 (324)	27 (187)

0.040-in (1.02-mm) sheet;
3-lobe Delta-Lok; diameter of
centerpunch = 0.25 in (6.4 mm)

10.2.4   Friction-fit joints

The extrusion process allows the designer to create shapes by snapping parts together (see examples in Fig. 10.25). In each case, the sections consist of two parts that snap together to form a box section. A question often considered by the designer is whether sufficient friction

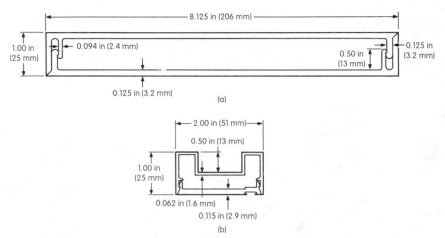

Figure 10.25 Sections for bending tests: (a) Section 2a, (b) Section 2b.

can be developed in the joints so that the parts behave as one section. Figures 10.26 and 10.27 are load-deformation results from tests of the two box sections pictured in Fig. 10.25. The behavior was similar for both cases. At very low loads the sections behaved as boxes. Afterward there was slip between the parts in shear, and the member behaved as two parts not attached together. Upon unloading, the members retained a bent shape. The loads at which slip occurred were less than 1 percent of the calculated failure loads assuming that the sections behaved as combined shapes.

Fully effective snap-lock joints in shear are not practical; additional fastening is needed. Tighter-fitting joints than those shown are possible, but assembly may be very difficult.

10.3 Adhesive Joints

Adhesives are used for secondary (not main load-carrying frame) parts of aircraft and automobiles, and in some primary parts of aircraft.[12] However, reliability of the manufacturing process is critical to achieve satisfactory joints, inspection methods to find deficient joints are not available, and long-term durability is questionable in some cases. Thus, although adhesives have great potential, they need to be carefully evaluated if the designer considers them as the only joining method for important structures.

Design information for the selection of adhesives and proportioning of some types of joints is available.[12–14] The designer should carefully study available information before specifying adhesive bonding for a critical structure. Some of the considerations follow.

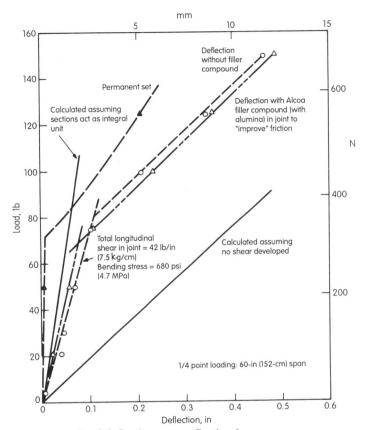

Figure 10.26   Load-deflection curves: Section 2a.

Design.   Closed-form solutions are available for stresses in lap joints for sheet construction, representative of aircraft construction. Design procedures for other types of joints (tubular, for example) and thicker materials are less defined.[13] Computer simulations are promising and useful for analyzing the behavior of the more complex joints. Ductile adhesives are available and should be employed for general-purpose applications. Tests are needed to verify behavior of most joints needed for primary structures.

Manufacturing.   The manufacturing process is critical to achieving good joints. Because inspection of joint quality is difficult, proper control of all steps of the process—i.e., surface preparation, application, and cure—is essential. There are mechanical and chemical treatments available for preparing the surface for adhesive bonding. Proper surface treatment is needed to ensure long-term durability

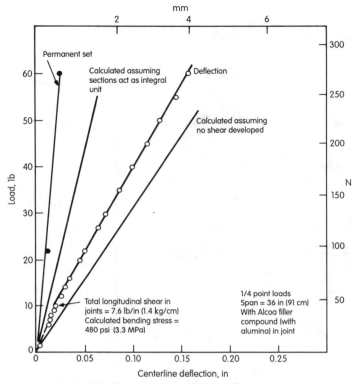

Figure 10.27   Load-deflection curves: Section 2b.

of the joint. Without proper treatment the interface is susceptible to corrosion, which will deteriorate the joint.

Performance.   Adhesive joints have the potential for outstanding performance, with very good static strength, stiffness, and fatigue strength compared to the other joining methods. The experience is largely with relatively thin aircraft structures; there is much less experience with thicker, general-purpose structures. The largest uncertainty with adhesive joints is their long-term durability. Work is under way to quantify some aspects of behavior.[14] More work is needed to quantify how well adhesive joints can sustain accidental damage and maintain structural integrity.

References

1. *1990 Structural Welding Code—Aluminum*, ANSI/AWS D1.2-90, American Welding Society, Miami, Florida.
2. *Welding Aluminum: Theory and Practice*, 1st ed., The Aluminum Association, Washington, D.C., March 1989.

3. "Specifications for Aluminum Structures," *Aluminum Construction Manual*, Sec. 1, The Aluminum Association, Washington, D.C., 1986.

4. Moore, R. L., Jombock, J. R., and Kelsey, R. A., "Strength of Welded Joints in Aluminum Alloy 6061-T6 Tubular Members," *Welding Journal*, April 1971.

5. Sharp, M. L., "Static and Dynamic Behavior of Welded Aluminum Beams," *Welding Journal Research Supplement*, February 1973.

6. Nelson, F. G., and Rolf, R. L., "Shear Strengths of Aluminum Alloy Fillet Welds," *Welding Journal Research Supplement*, February 1966.

7. Sharp, M. L., Rolf, R. L., Nordmark, G. E., and Clark, J. W., "Tests of Fillet Welds in Aluminum," *Welding Journal Research Supplement*, April 1982.

8. *Guidelines to Resistance Spot Welding Aluminum Automotive Sheet*, T10, The Aluminum Association, Washington, D.C. (undated).

9. Brungraber, R. J., "Strength of Welded Aluminum-Alloy Box Beams," *Welding Journal*, October 1960.

10. Dewalt, W. J., and Mack, R. E., "Design Considerations for Aluminum Fasteners," Paper No. 800455, presented at SAE Congress and Exposition, Society of Automotive Engineers, February 25–29, 1980.

11. Van Horn, Kent R. (ed.), *Aluminum*, Vol. 3, "Fabrication and Finishing," Chap. 11, American Society for Metals, Metals Park, Ohio, 1967.

12. Thrall, Edward W., and Shannon, Raymond W. (eds.), *Adhesive Bonding of Aluminum Alloys*, Marcel Dekker, New York, 1985.

13. Miller, J. M., Hammill, J. L., and Luyk, K. E., "Understanding Effects of Adhesive Ductility and Bondline Geometry on Tube-and-Socket Joint Performance," in Johnson, W. S. (ed.), *Adhesively Bonded Joints: Testing, Analysis, and Design*, ASTM STP 981, American Society for Testing and Materials, Philadelphia, 1988, pp. 252–263.

14. Nordmark, G. E., Miller, J. M., and Burleigh, D. T., "Environmental Durability of Joints for Aluminum Automotive Frames," NACE Paper 91566, presented at Corrosion 91, conference of the National Association of Corrosion Engineers, March 1991.

11

Fatigue

Fatigue is an important design consideration in most aluminum structures, and the most difficult problem to resolve. There have been estimates that up to 90 percent of the structural failures that occur in service are caused by fatigue. The reasons for the failures are that accurate load spectra and life prediction techniques generally are not available, particularly for new aluminum applications. The two problems are interrelated because even if the cyclic loads on the global structure are known the effects of these loads on the local stresses at a joint, essential for fatigue analysis, can be difficult to calculate. For example, the design techniques for many types of structures are two dimensional and determine nominal stresses in components only; thus they do not take into account three-dimensional behavior and resulting local stresses at attachments that are needed to estimate fatigue life. In addition, fabrication stresses and environment affect fatigue life, and thus they both need to be considered.

Despite the difficulties in designing for fatigue, aluminum is used successfully in many fatigue-critical structures, such as aircraft, ships, trucks, railroad cars, machinery, and highway and railroad bridges. Thus current techniques, when applied carefully, will overcome the difficulties and will result in satisfactory designs.

There are three general approaches to designing for fatigue: (1) the use of stress-cycle curves of typical joints and assembles,[1] (2) the use of crack propagation and fracture mechanics techniques,[2,3] and (3) the use of strain-controlled fatigue methods.[4,5] All these methods have merit and should be employed if the designers have adequate familiarity with the technique. The use of S-N curves is emphasized here because this design method is the most widely used of those mentioned. Also, in most non-aerospace structures, the design task is to proportion long-life structures that will not develop fatigue cracks; the S-N approach works well in this case.

In this chapter we discuss loadings, representative joint fatigue strengths, effects of spectrum loads, effects of environment, and other general issues.

11.1 Fatigue Loadings

Figure 11.1 shows schematics of several loading cases and definitions of some common terms. The upper three sketches show cyclic loading of constant amplitude, representative of most of the laboratory data available. Stress range, minimum and maximum stress, mean stress, and a stress ratio R are identified. The curves are the same except for mean stress, and fatigue performance is different in each case. The stresses are the combination of stresses from imposed loads plus those from fabrication. Residual stresses from welds in large, constrained

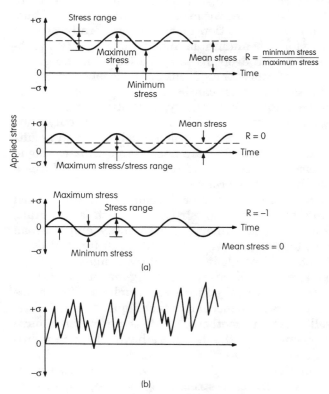

Figure 11.1 Loading spectra for fatigue. (a) Uniform cycles, (b) stress spectrum.

structures may be over one-half the yield strength of the material, and thus for welded construction stress range is often used for design. The effect of mean stress on fatigue strength is usually considered for mechanical fasteners. The effect of mean stress or stress ratio on fatigue strength of laboratory specimens is provided in the literature.[6] A method of predicting fatigue life under spectrum loading from constant-amplitude fatigue data is given in Sec. 11.4.

The lower sketch of Fig. 11.1 shows a spectrum that varies in amplitude and frequency, more representative of practical structures. Automotive and aerospace companies have measured spectra that they use for design and test verification of components and assemblies.

Figure 11.2 shows a stress spectrum that was estimated for the design of large, spherical tanks for transporting liquid natural gas. Aluminum was used for this application because of its excellent performance at cryogenic temperatures. The information included the distribution and number of cycles for stress (left side) and the order in which the loads are applied (daily, repeating cycle). Calculations and test verifications made use of these data. Note that a high, positive mean stress was estimated.

Figure 11.3 shows estimated spectra used for the design of the aluminum derrick for the Alcoa *Seaprobe* (described in Chap. 2). The values were estimated based on the cycles recorded for other similar ships operating in the North Atlantic Ocean and the anticipated roll and pitch characteristics of the *Seaprobe*. The derrick performed satisfactorily during service.

In some cases prototype structures are built and tested to obtain satisfactory load information. Figure 11.4 gives information obtained from the blades of a vertical-axis wind turbine in operation. The upper

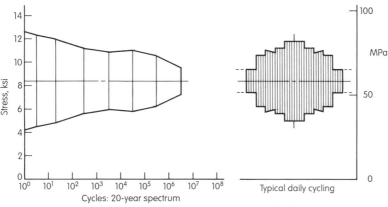

Figure 11.2 Test spectrum for liquefied natural gas tanks.[3]

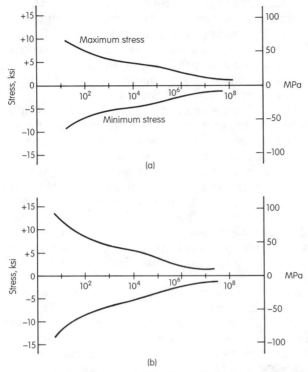

Figure 11.3 Spectra assumed for Alcoa *Seaprobe* derrick—20 years. (*a*) Roll motions, (*b*) pitch motions.

part of the figure is a partial recording of the strains at one point in the blade, and the lower part is a summary of how the blade stresses varied with wind speed. The stresses were difficult to calculate in this case because they included the effects of the wind load and the amplification of the loads because of dynamic behavior of the structure. Good spectra are needed for these machines because fatigue is the critical mode of failure.

Sometimes fatigue loadings are provided in codes and specifications. For example, information is given for highway structures[7] as required number of cycles at constant amplitude.

Many structures are designed and built without use of measured load spectra; light poles, overhead sign structures, and boats, for example. Past experience with designs that avoid problems is important. Also, it is essential to employ details that are good practice for fatigue-resistant joints.

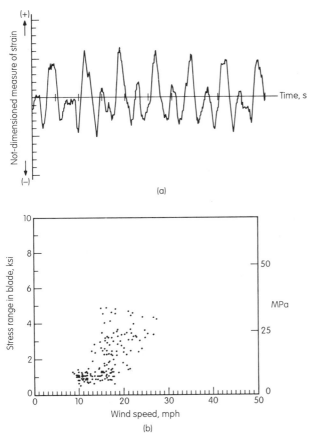

Figure 11.4 Stress data for aluminum blades on a vertical axis wind turbine. (a) Record of strain-gage readings. (b) Variation of blade stresses with wind speed.

11.2 Fatigue Strength of Typical Details

The procedure for designing new joints for fatigue strength using typical details is straightforward. The designer chooses a standard detail that should behave similarly to that of the new design and uses the fatigue strength that has been established for the standard detail. Designers experienced in fatigue behavior are usually very successful with this approach. In any critical design problem, this should be considered as a preliminary design; confirmation tests of assemblies representative of the actual structure must be made to finalize the design.

Various types of structural joints for which statistical minimums of fatigue strength have been developed are given in Fig. 11.5.[1] Additional explanation of these details is provided in Table 11.1 and the stress ranges for each category are shown in Table 11.2.[1] These values were obtained from an analysis of a large data base for aluminum joints, which has had input from many sources worldwide.[8] The information given is based on a confidence level of 90 percent and a probability that 95 percent of the data will exceed the specified values. There is no additional factor of safety applied. For simplicity in the specifications, step changes were allowed in Table 11.2. Figure 11.6 presents the data as originally developed in the form of straight-line segments. Except for mechanically fastened joints, the fatigue strengths are given as stress ranges, because high residual stresses are anticipated in most large, practical welded structures. The presence of large tensile residual stresses can result in fatigue cracks even if the externally imposed loads cause compressive cycles. The stresses given are nominal values and do not include any local effects. The fatigue strengths of mechanically fastened joints depend on stress ratio.

The fatigue strengths provided above are suitable for all alloys. There are some significant differences between alloys at the lower cycles, the higher-strength alloys having higher fatigue strengths, but there is not much difference at the higher cycles. One common problem in using the typical details arises when the new detail does not match any standard detail; then the fatigue strength is very uncertain.

Additional fatigue information on joints is provided here, but the data are limited, and therefore conservatism must be the rule if they are used for design. Figure 11.7 shows data for welded joints in a tubular truss.[9] The nominal axial stress was taken as the basis for plotting data, but the best interpretation of the data was obtained by using the most accurate calculation of the total applied stress at the edge of the weld, that is, the axial and bending stress in the member added to the stress from local bending of the tube wall. To obtain these stresses, a finite element analysis of the truss was made. The member stresses were used to solve for the local stresses at the location of fatigue cracks. No attempt was made to calculate the local stress concentration. The results of the calculations are given in Table 11.3. Figure 11.8[9] shows the location of failures.

Specimens shown in Fig. 11.9 simulating the shear plate and sill for a railroad car were tested in fatigue.[10] The sill dimensions are in Fig. 11.10, and some of the details at the end of the attachment of the shear plate to the sill are also given in Fig. 11.10. The results for the welded specimens are given in Fig. 11.11 and those for the riveted and rivbonded specimens are provided in Fig. 11.12. The finite element analyses of the joints showed that the local stresses near the points of

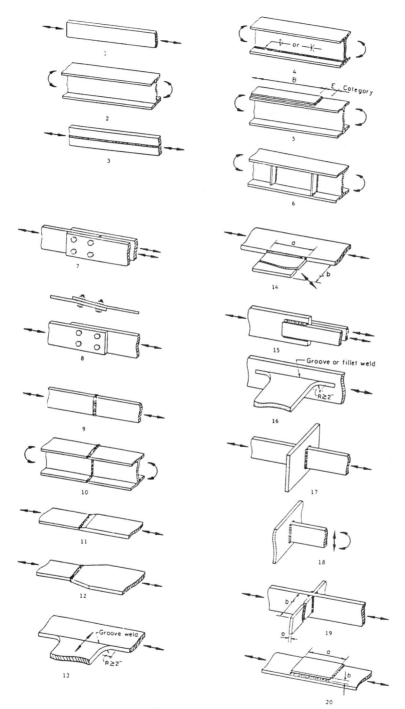

Figure 11.5 Typical details.[1]

TABLE 11.1 Stress Categories[1]

General condition	Situation	Stress category (see Table 11.2)*	Illustrative example nos. (see Fig. 11.5)†
Plain material, built-up members	Base metal with rolled or cleaned surfaces.	A	1, 2
	Base metal and weld metal in members, without attachments, built up of plates or shapes connected by continuous full- or partial-penetration groove welds or continuous fillet welds parallel to the direction of applied stress.	B	3, 4, 5
	Calculated flexural stress f_b in base metal at toe of welds on girder webs or flanges adjacent to welded transverse stiffeners.	C	6
	Base metal at end of partial-length welded cover plates having square or tapered ends, with or without welds across the ends.	E	5
Mechanically fastened	Base metal at net section of mechanically fastened joints which do *not* induce out-of-plane bending in connected material, where stress ratio, the ratio of minimum stress to maximum stress, SR is		
	$SR < 0$	C	7
	$0 \leq SR < 0.5$	D	7
	$0.5 \leq SR$	E	7
	Base metal at net section of mechanically fastened joints which induce out-of-plane bending in connected material.	E	8
Fillet-welded connections	Base metal at intermittent fillet welds.	E	
	Base metal at junction of axially loaded members with fillet-welded end connections. Welds shall be disposed about the axis of the members so as to balance weld stresses.	E	15, 17
	Weld metal of continuous or intermittent longitudinal or transverse fillet welds.	F	5, 15, 18

NOTE: See page 232 for footnotes.

TABLE 11.1   Stress Categories[1] (Continued)

General condition	Situation	Stress category (see Table 11.2)*	Illustrative example nos. (see Fig. 11.5)†
Groove welds	Base metal and weld metal at full-penetration groove-welded splices of parts or similar cross section ground flush, with grinding in the direction of applied stress and with weld soundness established by nondestructive inspection.	B	9
	Base metal and weld metal at full-penetration groove-welded splices at transitions in width or thickness, with welds ground to provide slopes no steeper than 1 to 2½, with grinding in the direction of applied stress, and with weld soundness established by nondestructive inspection.	B	11, 12
	Base metal and weld metal at full-penetration groove-welded splices, with or without transitions, having slopes no greater than 1 to 2½, when reinforcement is not removed and/or weld soundness is not established by nondestructive inspection.	C	9, 10, 11, 12
Attachments	Base-metal detail of any length attached by groove welds subject to transverse and/or longitudinal loading, when the detail embodies a transition radius, the radius of an attachment of the weld detail R, 2 in (51 mm) or greater, with the weld termination ground smooth:		
	$R \geq 24$ in (610 mm)	B	13
	24 in (610 mm) $> R \geq 6$ in (152 mm)	C	13
	6 in (152 mm) $> R \geq 2$ in (51 mm)	D	13
	Base metal at detail attached by groove welds or fillet welds subject to longitudinal loading, with transition radius, if any, less than 2 in (51 mm):		
	2 in (51 mm) $\leq a \leq 12b$ or 4 in (102 mm)	D	14
	$a > 12b$ or 4 in (102 mm)	E	14, 19, 20

TABLE 11.1 Stress Categories[1] (*Continued*)

General condition	Situation	Stress category (see Table 11.2)*	Illustrative example nos. (see Fig. 11.5)†
	where a = detail dimension parallel to the direction of stress, b = detail dimension normal to the direction of stress and the surface of the base metal.		
	Base metal at a detail of any length attached by fillet welds or partial-penetration groove welds in the direction parallel to the stress, when the detail embodies a transition radius R, 2 in (51 mm) or greater, with weld termination ground smooth:		
	$R \geq 24$ in (610 mm)	B	16
	24 in (610 mm) > $R \geq 6$ in (152 mm)	C	16
	6 in (152 mm) > $R \geq 2$ in (51 mm)	D	16
	Base metal at a detail attached by groove welds or fillet welds, where the detail dimension parallel to the direction of stress a is less than 2 in (51 mm).	C	19

*All stresses are T and Rev., where "T" signifies range in tensile stress only and "Rev." signifies a range involving reversal of tensile or compressive stress, except Category F, where stress range is in shear including shear stress reversal.

†These examples are provided as guidelines and are not intended to exclude other reasonably similar situations.

TABLE 11.2 Stress Ranges[1]

In ksi (MPa)

Stress category*	Design stress cycles			
	Up to 100,000	100,000–500,000	500,000–2,000,000	Over 2,000,000
A	17.5 (121)	13.6 (94)	11.0 (76)	9.5 (66)
B	16.0 (110)	11.4 (79)	8.5 (59)	6.0 (41)
C	12.0 (83)	8.2 (57)	5.9 (41)	4.0 (28)
D	9.5 (66)	6.4 (44)	4.5 (31)	3.0 (21)
E	7.5 (52)	4.7 (32)	3.2 (22)	2.0 (14)
F	6.0 (41)	3.8 (26)	2.5 (17)	1.6 (11)

*See Table 11.1.

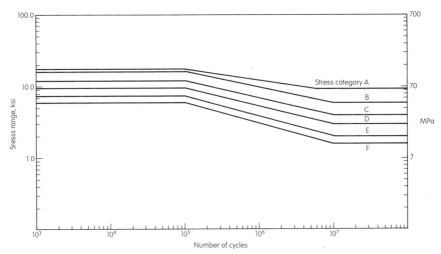

Figure 11.6   Stress range for typical details. (Stress categories are given in Table 11.1.)[1]

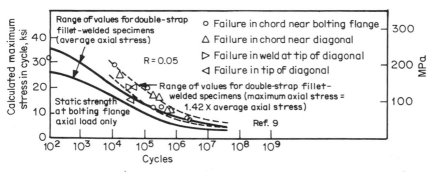

Figure 11.7   Fatigue of trusses considering bending of tube walls and members.[9]

TABLE 11.3   Ratio of Local Stress to Nominal Axial Stresses in Tubular Construction[9]

Member and location	Ratio: local to nominal
Chord member near bolting flange	1.97
Top of chord near diagonal	4.30
Tip of diagonal near tensile chord	3.04

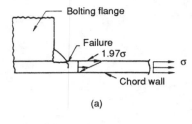

(a)

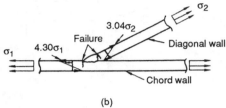

(b)

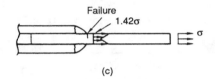

(c)

Figure 11.8 Local stress distribution in joints. (a) Bolting flange joint; (b) intersection of diagonal and chord; (c) double-strap fillet-welded specimen.[9]

fatigue failure (in the sill) were 1.7 to 2.1 times the nominal stresses used in the plots.

Fatigue results for spot welds in thin-sheet specimens are given in Fig. 11.13.[11] The results with several patterns of spot-welded joints are shown in Fig. 11.14. Similar data are presented for riveted joints (Fig. 11.15), mechanically clinched joints (Fig. 11.16), and adhesive joints (Fig. 11.17). All these tests are made with single fasteners in thin, single-lap specimens. Some of the clinches loosened during complete reversal tests as shown in Fig. 11.18, an undesirable behavior for a primary joining method.

Obviously, many other types of joints have been tested over the years, all of which give somewhat different results. While such data are helpful to the designer, a more general method for predicting the length of life of aluminum joints is needed.

11.3   Effects of Environment

The above information on fatigue considers behavior in a laboratory-type environment. Most practical structures are exposed to corrosive attack and to heat and cold. These effects are considered here; in cases where there will be elevated temperatures or possible severe corro-

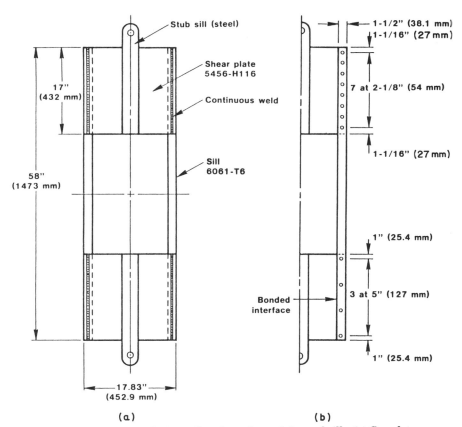

Figure 11.9 Specimens simulating railroad car shear plates and sills. (*a*) Complete specimen—welded. (*b*) One-half of specimen riveted-rivbonded.[10]

sion; the designer needs to incorporate them in addition to the other design considerations.

Corrosion. Figure 11.19 shows data for axial stress fatigue ($R = 0$) of smooth specimens, 0.125 in (3.2 mm) thick, in laboratory air and cyclic saltwater spray.[12] There is a significant reduction of fatigue strength at a million cycles due to the corrosion effects. There are differences depending on alloy, but the data base is insufficient to accurately define the differences between various alloys. Figure 11.20 gives fatigue data for smooth and mildly notched (central hole) specimens in flexure. The presence of the hole also reduces fatigue strength but not by as much as corrosion. Thus, corrosion has as much effect on smooth specimens in these tests as does a round hole. Some product designers have used fatigue strength of a sharply notched part for designing smooth parts, reportedly

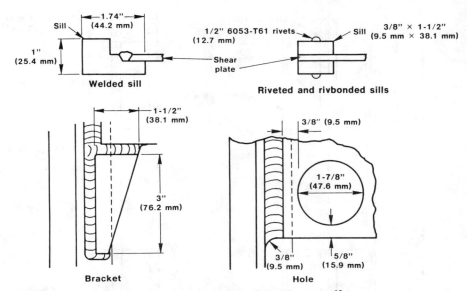

Figure 11.10 Details of specimens simulating railroad car joints.[10]

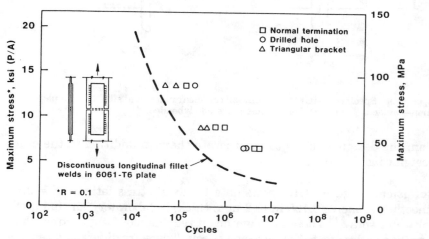

Figure 11.11 Fatigue strength of welded specimens simulating railroad car joints.[10]

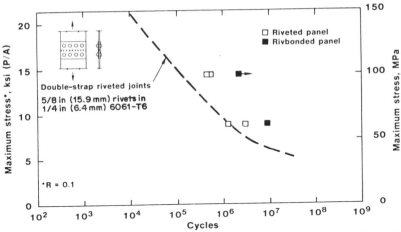

Figure 11.12 Fatigue strength of riveted and rivbonded specimens simulating railroad car joints.[10]

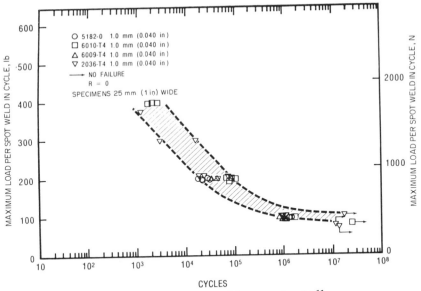

Figure 11.13 Fatigue strength of lap joints with one spot weld.[11]

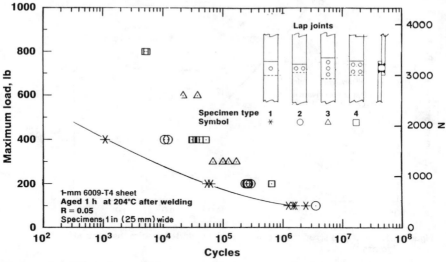

Figure 11.14   Effect of spot-weld pattern on fatigue strength of lap joints.

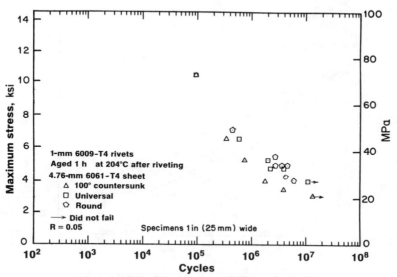

Figure 11.15   Effect of rivet type on fatigue strength of lap joints.

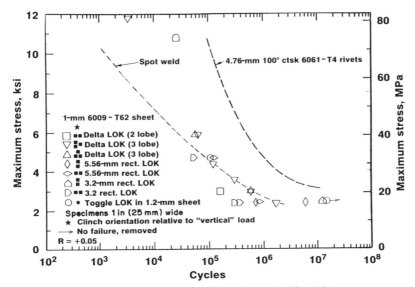

Figure 11.16   Effect of clinch type on fatigue strength of lap joints.

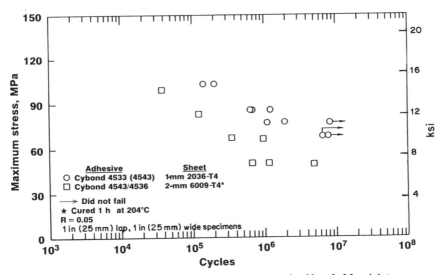

Figure 11.17   Effect of sheet thickness on fatigue strength of bonded lap joints.

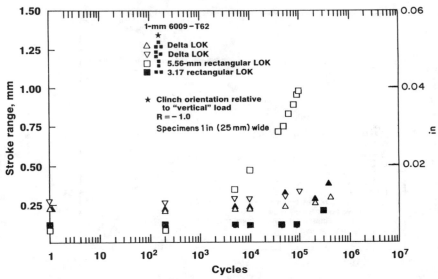

Figure 11.18 Slip of clinched joints under reversed loading of 445N.

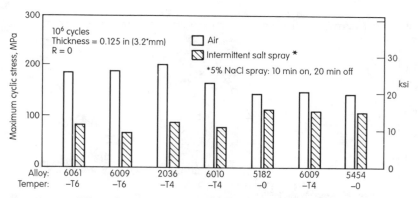

Figure 11.19 Effect of environment on axial fatigue strengths for aluminum auto body sheet alloys.[12]

because of their experience that the product designed for that strength performed adequately in long-term use in the environment. In retrospect, it appears that they are compensating for scratches, dings, and corrosion that the product will see during service by the use of fatigue data for sharply notched specimens.

The effects of corrosion are less for joints than for smooth specimens. Figures 11.21 and 11.22 show fatigue results for welded butt joints of 6061-T6 and 5083-H321, respectively, after various numbers of years

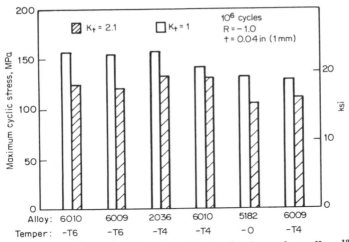

Figure 11.20 Flexural fatigue strengths for aluminum sheet alloys.[12]

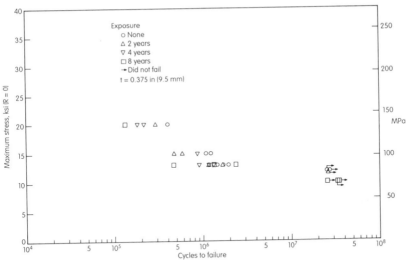

Figure 11.21 Effect of seacoast exposure on axial stress fatigue strength of 6061-T6 specimens with butt welds.[13]

of exposure at the seacoast. There is some trend to shorter lives of specimens with the longer exposure, but probably not enough to mandate a major change in design. Figure 11.23 shows that the bolted joints also were not affected significantly in fatigue by the seacoast exposure. Note that the specimens were subjected to exposure, then

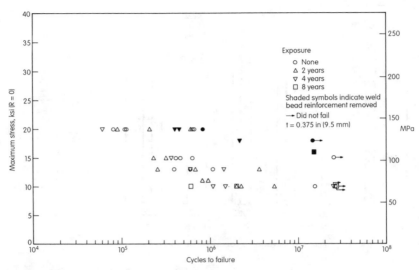

Figure 11.22 Effect of seacoast exposure on axial stress fatigue strength of 5083-H321 specimens with butt welds.[13]

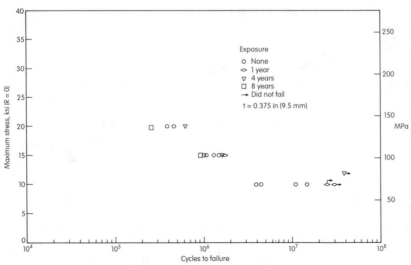

Figure 11.23 Effect of seacoast exposure on axial stress fatigue strength of bolted joints—7005-T63 plates with 2024-T4 bolts.[13]

tested. Usually designers do not take long-term exposure into account in the design of joints. This design philosophy probably is reasonable, but more research should be done to better understand the long-term durability of aluminum structures.

Temperature. Previous information presented in other chapters showed the beneficial effects of cryogenic temperatures on mechanical performance of aluminum parts. Table 11.4 gives data on fatigue.[14] Machined, round specimens of base metal and welded panels were utilized. Some sharply notched specimens are also included. The fatigue strengths at $-320°F$ ($-196°C$) were always higher than those at room temperature. It follows that the fatigue strength of structural joints should also be improved by low temperature.

Data for fatigue strength of alloys presented in Table 11.5 show that elevated temperatures reduce fatigue strength. Because results for both smooth and notched specimens show lower fatigue strength at elevated temperature than at room temperature, it would be expected that the fatigue strength of practical joints would also be lowered by elevated temperature.

The data presented are limited, and obtained under specific conditions, but they clearly indicate that temperature affects fatigue strength of aluminum and thus aluminum parts. The designer needs to conduct verification tests of any final design of a fatigue-critical structure.

TABLE 11.4 Effect of Low Temperature on Static and Fatigue Strength[14]

Base metal	Filler metal	Ratio: $\dfrac{\text{Property at } -320°F\ (-196°C)}{\text{Property at room temperature}}$				
		Tensile strength	Yield	Elonga-tion	Fatigue strength, cycles	
					10^5	10^6
5083-H113		1.33	1.17	1.49	1.15	1.18
5086-H32		1.38	1.18	1.62	1.18	1.15
5454-H32		1.42	1.17	1.72	1.15	1.25
5456-H321		1.31	1.15	1.42	1.12	1.14
5083-H113	5556	1.46	1.17	1.60	1.14	1.30
5086-H32	5356	1.37	1.06	1.06	1.21	1.32
5454-H32	5554	1.61	1.32	1.61	1.24	1.50
5456-H321	5556	1.32	1.16	1.12	1.17	1.30
5454-H32*	5554	1.34			1.80	
5456-H321*	5556	1.16			1.55	

Round specimens, 0.300 in (7.6 mm) diameter. Axial fatigue with a stress ratio of zero.
*Sharply notched specimens; $K_t = 19$.

TABLE 11.5 Fatigue Strength of Aluminum Alloys at Elevated Temperature*

				Ratio: $\dfrac{\text{Fatigue strength at elevated temperature}}{\text{Fatigue strength at room temperature}}$				
	Tempera-ture, °F				Fatigue strength at room temperature, cycles			
Alloy	(°C)	R†	K‡	Static	10^5	10^6	10^7	10^8
2024-T851§	300 (149)	0.5	1.0	0.84	0.93	0.92	0.89	0.84
			4.4	0.97	0.87	0.91	0.93	0.93
			>12	0.99	0.90	0.88	0.84	0.83
		0	1.0	0.84	0.85	0.83	0.77	0.72
			4.4	0.97	0.82	0.83	0.85	0.84
			>12	0.99	0.82	0.82	0.85	0.83
		−1	1.0	0.84	0.82	0.87	0.79	0.76
			4.4	0.97	0.78	0.81	0.85	0.83
			>12	0.99	0.90	0.91	0.89	0.88
5454-H34¶	300 (149)	−1	1.0		0.86	0.76	0.64	0.62
	400 (204)	−1	1.0		0.79	0.67	0.55	0.50

*Round specimens with diameter 0.5 in (12 mm) for axial fatigue, diameter 0.4 in (10.2 mm) for bending fatigue.
†Ratio minimum to maximum stress.
‡K = 1.0 is smooth specimen; K = 4.4 is circumferentially notched specimen with mild radius; $K \geq 12$ is circumferentially notched specimen with sharp notch.
§Axial fatigue.
¶Bending fatigue.

11.4 Life Prediction with Stress Spectrum

The most common tool for estimating the fatigue life of a structure under a spectrum of loadings is Miner's linear cumulative damage rule.[15] This rule estimates fatigue life by the following expression:

$$N_g = \frac{1}{\displaystyle\sum_{i=1}^{j} \frac{\alpha_i}{N_i}} \tag{11.1}$$

where N_g = fatigue life under spectrum of loads
α_i = fraction of fatigue life for each stress level, σ_i
N_i = fatigue life at constant stress amplitude for stress level, σ_i

The application of this rule is described in Fig. 11.24. The stress spectrum in terms of number of cycles at each stress level and the S-N curve for the particular joint and stress ratio are required. Studies of the accuracy of this rule[16] show that it can be conservative or unconservative, perhaps by a factor of 2 on life, depending on joint detail and load spectrum. However, this accuracy is probably sufficient

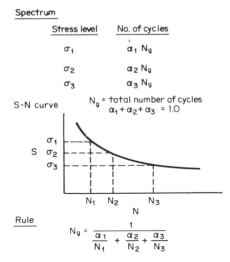

Figure 11.24 Miner's cumulative damage rule.

for preliminary design, with test confirmation needed for fatigue-critical final designs.

11.5 Other Observations on Fatigue

11.5.1 Effect of joining method on fatigue

The brief study that follows of welded and riveted (bolted) joints makes use of the data presented for 6061-T6[6]; the information is applicable to other alloys since long-life fatigue strength of joints is relatively independent of alloy. The comparisons are based on gross stress on the section, reflecting the load-carrying capacity. In addition, curves representing lower bounds to the test data are utilized. Figure 11.25 shows that double-strap riveted joints have a fatigue strength similar to welded butt joints. However, the fatigue strength of riveted joints continues to decrease at long lives, whereas that for the butt welds bottoms out.

Riveted joints are much better in fatigue than fillet-welded joints, as shown in Fig. 11.26. The same conclusion is reached based on fatigue tests of beam splices given in Fig. 11.27. Thus, riveted construction can provide better fatigue performance than welded construction, except for those structures in which butt welds only may be used. Figure 11.28 shows that, generally, riveted and bolted joints have about the same fatigue performance.

Riveted joints can have low fatigue strengths at long lives due to fretting failures. *Fretting* is effectively a local friction weld between the adjoining plates in the joint due to the small sliding movements

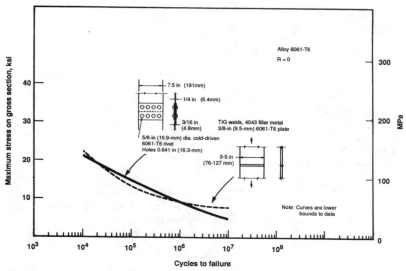

Figure 11.25   Butt weld and rivet joint fatigue tests.[6]

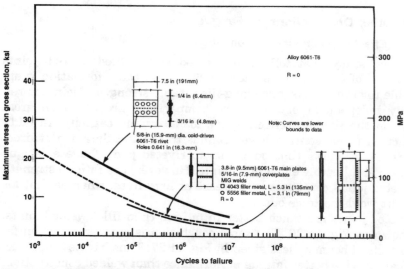

Figure 11.26   Fillet-welded and riveted joint tests.[6]

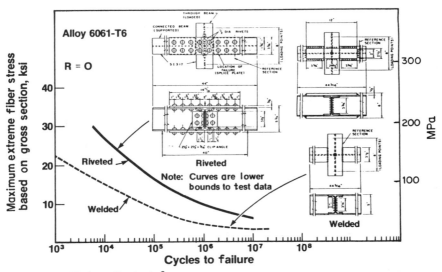

Figure 11.27   Beam splice tests.[6]

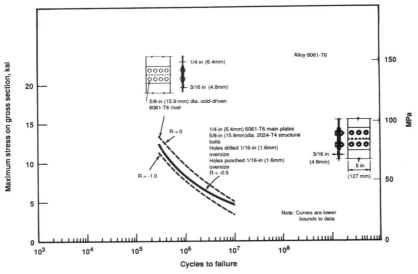

Figure 11.28   Riveted and bolted joint tests.[6]

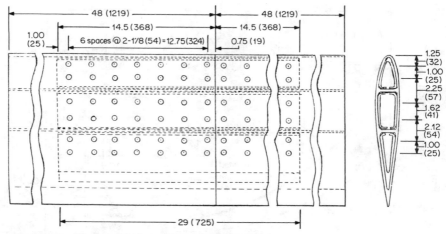

Figure 11.29 Blade-to-blade specimen for fatigue test.

that take place in the joint. Figure 11.29 shows a riveted splice joint evaluated for the blade of a vertical-axis wind turbine. The results of the fatigue tests of the joint (in bending) given in Fig. 11.30 show that fatigue strength continues to drop with no indication that it will bottom out, characteristic of fretting. Mechanically fastened structures

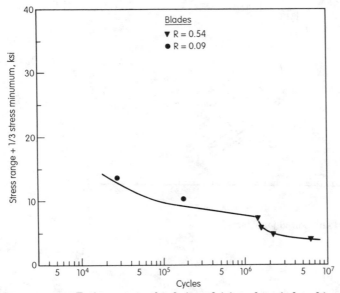

Figure 11.30 Fatigue strength of riveted joints for wind turbine blades.

can and have developed fatigue cracks with very low cyclic loads, causing considerable surprise to the designer. Adhesives in the joint, if carefully applied, can prohibit fretting if there are no starved areas in which the plates can have contact. The ordinate in Fig. 11.30 is the stress range plus one-third of the minimum stress, used to remove most of the mean stress effects from the data.[6]

Adhesive joints or joints with adhesives in combination with other joining methods usually will have much better fatigue behavior than welded or mechanical joints. However, their use is limited because of the other factors discussed in Chap. 10.

11.5.2 Effect of alloy

There is a significant effect of alloy on fatigue strength of smooth specimens and joints at low lives. Figure 11.31 shows results for butt welds. A common curve is often employed for structural design of joints because there is little difference in strength at long lives.

11.5.3 Effect of local stress

The S-N curves are usually based on nominal stress in the component. The local stress at the joint, however, is a critical factor in the fatigue strength. Usually it is helpful to do a finite element analysis of the structure, similar to that described above for the tubular truss, to obtain these local stresses. Even if loads are not known accurately, it is worthwhile to minimize the local stresses.

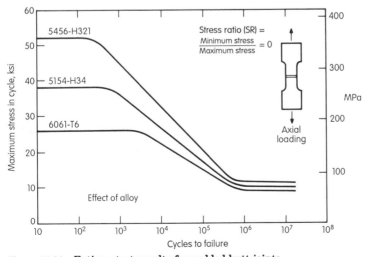

Figure 11.31 Fatigue test results for welded butt joints.

In a number of the discussions above, it has been stated that a verification test should be done, using an assembly that closely simulates the actual structure being designed. This step should be considered for all new, important design problems in which fatigue is critical.

References

1. *"Specifications for Aluminum Structures," Aluminum Construction Manual,* Sec. 1, The Aluminum Association, Washington, D.C., 1986.
2. Kaufman, J. G., Bucci, R. J., and Kelsey, R. A., "Fracture Mechanics Aspects of the Structural Integrity Technology of Spherical Aluminum Containment Vessels for LNG Tankers," *Journal of Engineering Materials and Technology,* July 1980.
3. Kelsey, R. A., Nordmark, G. E., and Clark, J. W., "Fatigue Crack Growth in Aluminum Alloy 5083-0 Thick Plate and Welds for Liquefied Natural Gas Tanks," Special Technical Publication 556, American Society for Testing and Materials, Philadelphia, 1974.
4. Wong, W. A., Bucci, R. J., Stentz, R. H., and Conway, J. B., "Tensile and Strain-Controlled Fatigue Data for Certain Aluminum Alloys for Application in the Transportation Industry," presented at the 1987 SAE International Congress and Exposition, Society of Automotive Engineers, Detroit, Michigan, February 23–27, 1987.
5. Landgraf, R. W., and LaPointe, N. R., "Cyclic-Stress-Strain Concepts Applied to Component Life Prediction," SAE Paper 740280, 1974.
6. Nordmark, G. E., and Clark, J. W., "Fatigue of Joints in Aluminum Alloy 6061-T6," *Proceedings of the American Society of Civil Engineers,* Journal of the Structural Division, December 1964.
7. *Standard Specifications for Highway Bridges,* 13th ed., The American Association of State Highway and Transportation Officials, Washington, D.C., 1983.
8. Sanders, W. W., Ondra, R., and Kosteas, D., "State-of-the-Art of the ALFABET Project," paper presented at 5th INALCO '92, International Conference on Aluminum Weldments, Munich, April 27–29, 1992.
9. Sharp, M. L., and Nordmark, G. E., "Fatigue Strength of Welded Tubular Aluminum Truss," *Proceedings of the American Society of Civil Engineers,* Journal of the Structural Division, August 1977.
10. Nordmark, G. E., and Sharp, M. L., "Fatigue Tests of Shear Plate, Side Sill Joints for Railway Cars," presented at the Winter Annual Meeting, American Society of Mechanical Engineers, Miami Beach, Florida, November 17–21, 1985.
11. Nordmark, G. E., "Fatigue Performance of Aluminum Joints for Automotive Applications," SAE Paper 780397, Society of Automotive Engineers, Warrendale, Pennsylvania, 1978.
12. Bucci, R. J., Herbein, W. C., and Mueller, L. N., "Representative Mechanical and Fatigue Properties of Autobody Sheet Alloys," addendum to presentation, "Selecting Aluminum Alloys to Resist Failure by Fatigue and Fracture Mechanisms," presented at SAE Congress and Exposition, Detroit, Michigan, February 25–29, 1980.
13. Nordmark, G. E., and Kelsey, R. A., "Fatigue Tests of Weathered Aluminum Bolted and Welded Joints," First International Aluminum Welding Conference, American Welding Society, Cleveland, Ohio, April 8, 1981.
14. Kaufman, J. G., and Nelsen, F. N., "Cryogenic Temperatures Up Fatigue Strengths of Al-Mg Alloys," *Space/Aeronautics,* July 1962.
15. Heywood, R. B., *Designing against Fatigue of Metals,* Reinhold, New York, 1962.
16. Gurney, T. R., "Fatigue Tests under Variable Amplitude Loading," Research Report 220/1983, The Welding Institute, Cambridge, England, July 1983.

Special Design Problems

This chapter covers several design problems: vibration of members and structures, energy absorption, and general toughness issues. All have particular importance for some types of aluminum structures. The discussion of vibration is primarily concerned with wind-induced vibration. Designing for energy absorption can be to maximize or to minimize the value to failure. Toughness is discussed as a combined material and geometry problem. The final section reviews potential corrosion problems and their prevention.

12.1 Vibration

Vibration of members, assemblies, and global structures will generally lead to problems, usually fatigue failures, and thus it needs to be controlled or eliminated. Vibrations are mechanically induced in structures by suddenly applied loads and cyclically applied loads from operating machinery. Light poles and overhead sign structures mounted on vibrating bridges have experienced severe dynamic loads. Vibration often occurs in structures subjected to wind or other moving fluids because of the associated periodic forces from the fluid as it flows over the structure.

Large amplitudes of vibration can result from periodic forces that are close to a natural frequency of the structure, particularly with the modern, lightly damped welded construction. The resulting stresses can be large and cause premature failure.

Large amplitudes usually will not occur if the frequency of parts, assemblies, and the structure lies outside a range from one-half to twice the frequency of any regularly repeated impulses. The natural frequency of a member or assembly may be calculated by means of the formulas given in Table 12.1.[1] In the case of members in the horizon-

TABLE 12.1   Natural Frequencies of Vibration—Beams or Springs[1]

$$f = \frac{K_1}{\sqrt{D}}$$

where f = natural frequency of vibration, Hz

D = maximum static deflection of member under its own
weight plus any weights which vibrate with it, in

K_1 = factor below

Case		Description	K_1 Deflection D, in (mm)
1		Concentrated weight on relatively light beam or spring	3.13 (15.8)
2		Uniformly distributed weight on beam simply supported at ends	3.55 (17.9)
3		Uniformly distributed weight on beam fixed at ends	3.55 (17.9)
4		Uniformly distributed weight on cantilever beam	3.89 (19.6)

tal position, the deflection is calculated in the ordinary manner; in the case of members whose position is other than horizontal, the deflection is still calculated as though the member were in the horizontal position. This use of the deflection is simply a convenient method of taking into account the stiffness and span length of the member and the magnitude and distribution of the mass that is in motion when the member is vibrating. The equations are accurate for trusses and other assemblies as well as for simple members.

Table 12.1 gives first-mode frequencies of usual interest in vibration analysis. Note that higher modes can be excited if their frequency of vibration nearly synchronizes with the frequency of some repeated impulses, and thus these higher modes may have to be considered.

12.1.1 Individual members

The methods of avoiding or limiting vibration of individual members are by proper proportioning of the member, or by the addition of structural damping. The following discussion shows the effectiveness of damping tapes (strips of aluminum foil bonded to the member by an energy-absorbing adhesive) applied to vibrating tubes and the ineffectiveness of polyurethane foam in the tube for providing damping. The setup for the evaluation is given in Fig. 12.1. Thin flexure plates were employed at the ends to achieve a simply supported condition with low damping. Load-deformation results are shown in Fig. 12.2 for empty and foam-filled (polyurethane) tubes. The free vibration responses of the various cases tested are given in Fig. 12.3. There is low damping in the empty and foam-filled tubes. The differences in damping at the higher stresses reflect differences in the specimens. Significant damping is achieved by the tape, increasing as the amount of tape is increased.

Slender tubes of overhead sign trusses and angles of transmission towers have failed in fatigue because of wind-induced vibration. The most

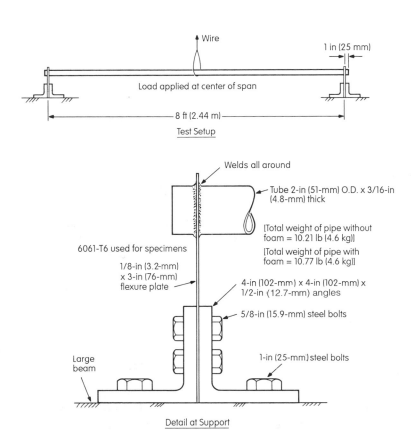

Figure 12.1   Setup for evaluating damping tape.

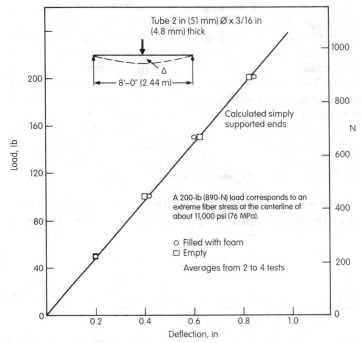

Figure 12.2 Load-deformation curves of tubular beam specimens. (Specimens shown in Fig. 12.1.)

common way to control the vibration is to properly proportion the members. The following paragraphs provide some guidance for design.

In the case of tubular members, the cause of vibration is well known; it occurs because of the regular transverse impulses to the member from shedding of vortices as the wind flows over the member. The approximate frequency of shedding of these vortices is as follows:[2]

$$f = \frac{SV}{d} \tag{12.1}$$

where f = frequency of vortex shedding, Hz
V = wind velocity, mph (kph)
d = outside diameter of tube, in (mm)
S = approximate Strouhal number for tubes, 3.26 for mph/in and 51.5 for kph/mm units

Vibration occurs if the natural frequency of the tube is close to that for vortex shedding. In this event, flexural vibration occurs in a direction perpendicular to that of the wind. Equation (12.1) may be rear-

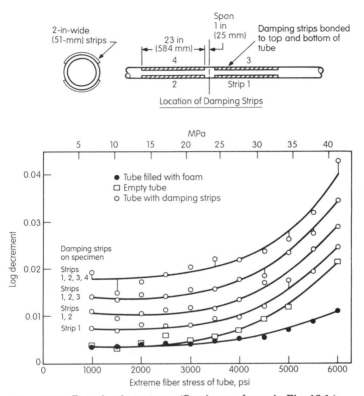

Figure 12.3 Damping in systems. (Specimens shown in Fig. 12.1.)

ranged to calculate the velocity of wind that will cause vibration in the member:

$$V = \frac{fd}{S} \tag{12.2}$$

In Eq. (12.2), f is the natural frequency (but not always the first mode) of the member's vibration. Vibration can occur for a range of wind velocities in the vicinity of that given by Eq. (12.2), so members subject to steady winds should be proportioned so that the velocity given by this equation is well outside (perhaps twice) the maximum expected wind velocity.

The behavior of open shapes, such as angles, is not as well understood as that for cylinders, but vibration does occur depending on the angle of impingement of the wind, the proportions of the shape, and the velocity of the wind. In these shapes, torsional vibrations are of concern, although flexural vibrations may also occur. The natural frequency of vibration for a member in torsion can be calculated by the use of the equivalent slenderness defined previously for torsional buckling. The equation is as follows:

$$f = \frac{C_v}{\lambda_t L} \qquad (12.3)$$

where f = first-mode frequency of torsion, Hz
$\quad L$ = length of member, in (mm)
$\quad C_v$ = 3.1 · 10⁵ for in units and 78.7 · 10⁵ for mm units
$\quad \lambda_t$ = equivalent slenderness ratio for torsional buckling ≅ 5.13
$\quad \sqrt{I_p/J}$
$\quad I_p = I_x + I_y$ = sum of moments of inertia about shear center
$\quad J$ = torsion constant

Equation (12.3) is a fairly general formula for both torsional and flexural vibration of members supported at both ends. For flexural vibration the equation becomes:

Beams with hinged supports: $\quad f = \dfrac{C_v}{\lambda_f L} \qquad (12.4)$

where λ_f = slenderness ratio of member as column in flexural buckling
$\quad C_v$ = 3.1 · 10⁵ for in units, 78.7 · 10⁵ for mm units
$\quad f$ = first-mode natural frequency in bending

Beams with fixed supports: $\quad f = \dfrac{C_v}{\lambda_f L} \qquad (12.5)$

where C_v = 3.6 · 10⁵ for in units, 91.4 · 10⁵ for mm units.
    Introducing Eq. (12.3) or Eq. (12.4) into Eq. (12.2) gives the following relationship between the proportions of the member and the wind speed that will cause vibrations:

$$V = \frac{0.95 \cdot 10^5}{\lambda(L/d)} \qquad (12.6)$$

where λ = slenderness ratio for flexural or torsional buckling.
$\quad V$ = wind velocity, mph

A number of aluminum shapes have been tested in a wind tunnel using the setup shown in Fig. 12.4. The types of members are given in Table 12.2 along with other members described in the literature[3] and some field observations of members vibrating in the wind. All the data are plotted in Fig. 12.5 with wind velocities at which vibration occurred and the effective slenderness of the member or assembly. Equa-

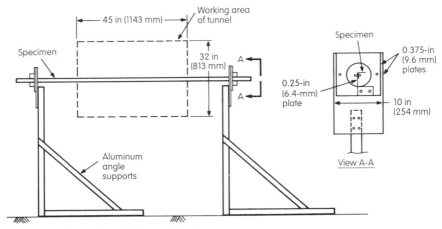

Figure 12.4 Setup for wind tunnel tests.

tion (12.6) gives a reasonable trend for the lower velocities. A better lower bound to the data is given by the following equation, which gives a wind velocity about two-thirds that of Eq. (12.6):

$$V = \frac{0.6 \cdot 10^5}{\lambda(L/d)} \qquad (12.7)$$

where V = wind velocity in mph.

The data plotted in Fig. 12.5 are the most severe cases recorded for the tests. Some specimens did not vibrate for some of the angles of impingement and wind velocities. Vibration at lower velocities was in the first mode and the amplitude increased with wind velocity. Higher modes were excited at the higher wind velocities. The data clearly show that flexible members, in torsion and bending, are susceptible to wind-induced vibration.

In order to make use of this information, a wind velocity must be defined. Vibration problems have generally occurred where structures are subjected to lower-velocity winds that are more uniform and steady in nature than the higher winds. Table 12.3 gives some very approximate suggestions based on limited experience. The designer should also make use of information about other designs that are performing satisfactorily in service.

12.1.2 Vibration of overhead sign structures

Tubular overhead sign trusses have vibrated in mild winds before signs were installed, resulting in fatigue cracks.[2] Analysis of the

TABLE 12.2 List of Specimens Subjected to Wind-Induced Vibration

Alcoa's wind tunnel tests*

Specimen no.	Member	Shape	Span, in
1	∠ 1½ × 1½ × 3/32	(single angle)	75
2	2∠s 1½ × 1½ × 3/32	(T)	75
3	2∠s 1½ × 1½ × 3/16	(T)	75
4	2∠s 1½ × 1½ × 3/32	(T-L)	75
5	2[s 1¼ × 5/8 × 1/8	(double channel)	75
11	∠ 1 × 1 × 1/16 (steel)	(single angle)	75
12	Same as No. 2		100
13	Same as No. 2		50
14	Same as No. 2		100
15	Same as No. 2		75
16	Same as No. 2		100
17	Same as No. 4		100
18	Same as No. 4		50
19	2∠s 1½ × 1½ × 3/32	(T)	100
20	Same as No. 19		75
21	Same as No. 19		50
22	2∠s 1 × 1 × 1/16	(T)	50
23	Same as No. 1		100
24	Same as No. 1		50
25	4∠s 1½ × 1½ × 3/32	(cross)	75

Canadian wind tunnel tests† (Ref. 3)

Specimen no.	Member	Shape	Span, in
6	2∠s 2½ × 1¾ × 0.12	(T)	204
7	2∠s 3 × 2 × 0.19	(T)	204
8	2∠s 2 × 2 × 1/8 (steel)	(T)	204
9	2∠s 2½ × 1¾ × 0.12	(T-L)	204
10	∠ 2 × 2 × 0.10	(single angle)	204

Field observations, tubes

Specimen no.	Member	Span, in
27	6 in φ × 3/16	192‡
26	1¾ φ × 1/8	116§

Conversion: 1 in = 25.4 mm. *Wind velocities considered: 0 to 45 mph. †Wind velocities considered: 30 to 140 mph. ‡The ends were flattened and bolted to gusset plates. §The ends were welded to tubular chord members, 4½-in O.D. × 3/16-in wall.

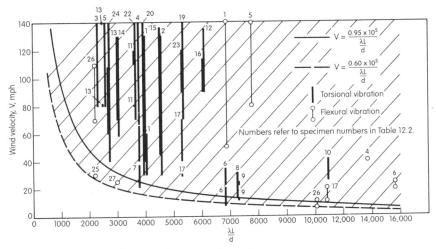

Figure 12.5 Aeolian vibration of structural members.

power input from the wind and of natural damping in these structures, illustrated in Fig. 12.6, shows that the power that is dissipated
by natural damping is low compared to that input by the wind. The
amplitude of vibration increases until the power input equals the dissipated power (amplitude A in Fig. 12.6). Additional damping must be
applied to prevent vibration. The upper curve in Fig. 12.6 shows the
effect of a damper efficient enough to prevent the vibration.

The power input to the truss by the wind is provided by the following equation.[2] The calculated value is based on the assumption that
the four main chords are entirely responsible for the vibration.

$$P = 1.14 \cdot 10^{-4} \frac{\rho}{g} L d^4 f^3$$

$$\times \left[2{,}220 \left(\frac{y}{d} \right)^2 - 13{,}100 \left(\frac{y}{d} \right)^3 + 36{,}300 \left(\frac{y}{d} \right)^4 \right] \quad (12.8)$$

where P = power input, in $\cdot$ lb/s (mm $\cdot$ kg/s)
$\quad$ d = diameter of chord, in (mm)
$\quad$ L = truss span, in (mm)
$\quad$ y = single amplitude of vibration, in (mm)
$\quad$ f = natural frequency of vibration of structure, cps (Hz)
$\quad$ g = acceleration of gravity, in/s^2 (mm/s^2)
$\quad$ ρ = unit weight, lb/in^3 (kg/mm^3)

Figure 12.7 shows that calculated values from Eq. (12.8) are conservative (larger) compared to values measured on a 78-ft (23.8-m)

TABLE 12.3   Tentative Requirements for Structural Elements for Aeolian Vibration

Type of member	Type of vibration	Limiting proportions, $V = 20$ mph*†	Limiting proportions, $V = 30$ mph†‡
	Flexural	$\dfrac{L}{r} \le 95$	$\dfrac{L}{r} \le 77$
	Flexural	$\dfrac{L}{r} \le 103$	$\dfrac{L}{r} \le 84$
	Torsional	$\dfrac{L}{t} \le 410$	$\dfrac{L}{t} \le 270$
	Flexural	$\dfrac{L}{r} \le 100$	$\dfrac{L}{r} \le 82$
	Torsional	$\dfrac{L}{t} \le 580$	$\dfrac{L}{t} \le 390$

*Maximum uniform wind assumed for most locations with rolling terrain, trees and vegetation, and buildings.
†Based upon Eq. (12.7), assuming $K = 1.0$ for flexural vibration and average values of radius of gyration of shapes.
‡Maximum uniform wind assumed for areas with flat, open terrain such as over water.

welded tubular truss undergoing wind-induced vibration. If a three-chord truss is used, the values for power input should be about three-fourths of those given by Eq. (12.8).

A practical way of preventing the vibration in these trusses is to apply external damping. Three dampers shown in Fig. 12.8 have been found to be useful. Their characteristics are given in Fig. 12.9. The smaller damper is efficient for higher frequencies (usually short spans) and the larger damper is efficient for the lower frequencies (usually long spans). To determine the effectiveness of a damper in preventing vibration, the power input from the wind is calculated using Eq. (12.8) and plotted as shown in Figs. 12.6 or 12.7. The power dissipated at several amplitudes by one of the dampers is taken from Fig. 12.9 and placed on the same plot. If the power dissipated is higher than that input by the wind, the truss does not vibrate.

Usually one damper is sufficient. If more than one damper is used, the power dissipated is the sum of the values from all the dampers. The natural damping of the truss is best neglected because it is low compared to that of the damper and cannot be estimated accurately.

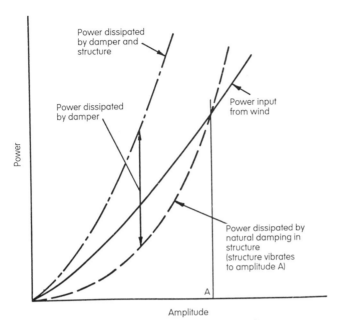

Figure 12.6   **Analysis for damping requirements.**[2]

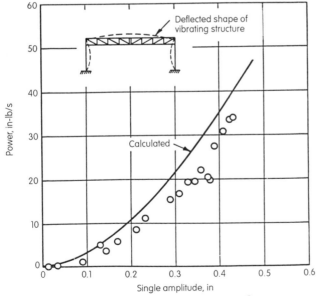

Figure 12.7   **Power input from wind to structure.**[2]

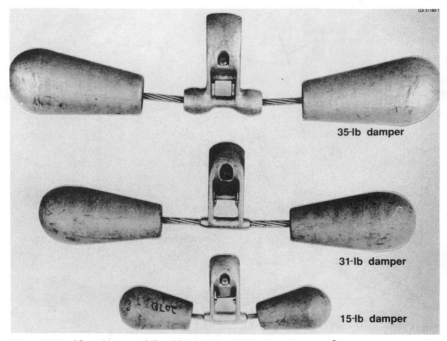

Figure 12.8   Alcoa improved Stockbridge-type vibration dampers.[2]

First-mode vibration usually occurs in the truss, and thus the most efficient location for the damper is at midspan.

Depending on their geometry, some light poles and flagpoles can vibrate excessively in the wind. Manufacturers of these products know from experience what types of products can have problems and they use dampers that prevent the vibration.

12.2   Energy Absorption

The amount of energy that is absorbed by a deforming structure is important in some design problems. An efficient system that absorbs a large amount of energy is needed for the front structure of an automobile, for example. The base or a device at the base of a light pole must be strong for wind loads, but break away readily under a car impact (low energy). The amount of energy that the structure can absorb before failure is highly dependent upon the design of the structure.

Figure 12.10 presents the energy per volume of metal developed by different members of 6061-T6 under axial or bending loads. This alloy is widely used and generally considered to be tough and good for general-purpose use. The energy under the stress-strain curve was

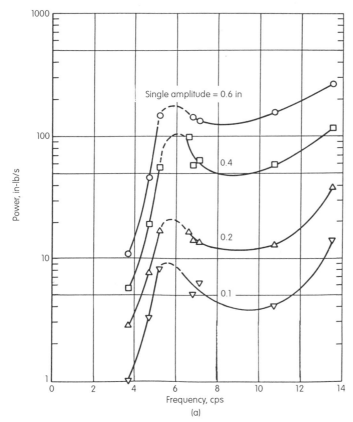

Power, in-lb/s

Frequency, cps

(a)

Single amplitude = 0.6 in

0.4

0.2

0.1

Figure 12.9 Power dissipated by dampers: (*a*) 15-lb damper, (*b*) 31-lb damper, (*c*) 35-lb damper.[2]

used to calculate the tensile test results shown. The results of crush tests of tubes have a range of values depending on their proportions. Results for the simply supported beams have been reported previously.[4] The cantilever beams had different shapes and joints at the base. Only the volume of the cantilever beam was used in the calculations; volumes of fasteners, support elements, etc. were neglected. With all specimens, significant inelastic deformation occurred. The elastic energy is relatively small compared to the total energy shown; elastic energy is calculated to be less than 2 percent of the total energy for the axial tensile test.

There are some observations for design. The tensile and compressive specimens had the highest energy because all or almost all of the material is inelastically deformed. The crush specimens have a higher

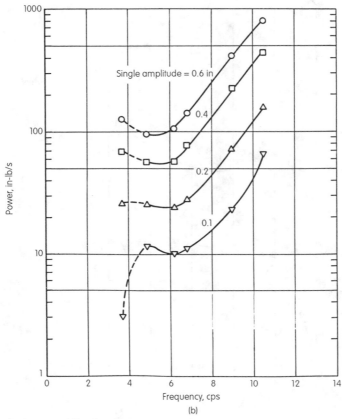

Figure 12.9 (*Continued*)

energy than the tensile specimen, apparently because necking in the tensile member limits energy, but necking does not occur in the folding of the compressive member. Joints may also limit energy absorption in practical tensile members, for example.

All the bending members had lower energy absorptions than those for the axially loaded members because the inelastic deformation occurs in the beams only at high-stress locations. Thus, only a small portion of the span and only part of the cross section are inelastically deformed. Simply supported beams with no welds or with longitudinal welds had about the same energy to failure. The lowest energy occurs for the beam with a transverse butt weld at the point of highest stress (midspan). Because the strength of the material near the weld is lower than the yield strength of the base material, the inelastic deformation occurs in a small volume of material only, and the energy is low. The

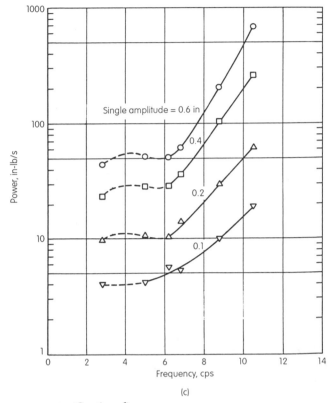

Figure 12.9 (*Continued*)

results for the other beams vary because of the fastener designs. Low strength of the fasteners or in the joint, compared to the member yield strength, results in low energies.

Note that even with a "tough" alloy, energy absorption is highly dependent upon design. Some guidelines for designing for high energy absorption follow:

- Design joints to have a strength equal to that of the members. Yielding needs to occur in the members as well as the joints.

- Locate joints in low-stress areas of the structure, particularly if they cannot be designed for full member strength.

- Stress the structure as uniformly as possible so that most of the volume is inelastically deformed at failure.

- Employ axial crush members.

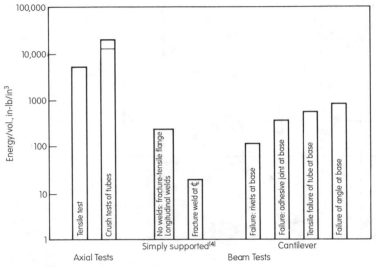

Figure 12.10   Energy absorption in 6061-T6 members.

For design problems in which the aim is to minimize energy absorption, the converse of each of the above is employed.

Crush behavior is important in energy management in automobiles, and considerable work has been done to determine behavior. For the axial crush of thin-walled aluminum shapes, the following equation has usefulness:

$$P_m = C_e t^{5/3} b^{1/3} \sigma_y^{2/3} E^{1/3} \tag{12.9}$$

where P_m = mean load in crush
  C_e = coefficient depending on shape and speed of loading
  t = thickness of part
  b = average of height and width dimensions of cross section
  σ_y = compressive yield strength
  E = modulus of elasticity

Some definition of terms and behavior are needed. Figures 12.11 and 12.12 show crushed tubes. The appearance is different because of the difference in proportions of the tubes, but all specimens were efficient in energy absorption. Also, note that there have been large inelastic deformations of the material, but little fracture of the material, and that these are conditions necessary for good energy absorption. A load-deformation plot is given in Fig. 12.13. The initial peak in the curve is highest and corresponds to the first fold that develops; subse-

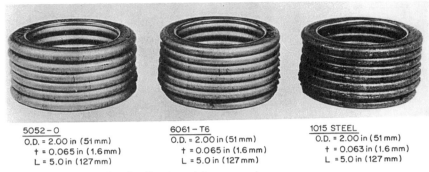

5052 - O
O.D. = 2.00 in (51 mm)
t = 0.065 in (1.6 mm)
L = 5.0 in (127 mm)

6061 - T6
O.D. = 2.00 in (51 mm)
t = 0.065 in (1.6 mm)
L = 5.0 in (127 mm)

1015 STEEL
O.D. = 2.00 in (51 mm)
t = 0.063 in (1.6 mm)
L = 5.0 in (127 mm)

Figure 12.11 Tubes after loading in axial compression.

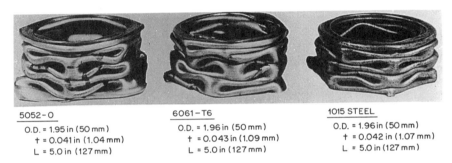

5052 - O
O.D. = 1.95 in (50 mm)
t = 0.041 in (1.04 mm)
L = 5.0 in (127 mm)

6061 - T6
O.D. = 1.96 in (50 mm)
t = 0.043 in (1.09 mm)
L = 5.0 in (127 mm)

1015 STEEL
O.D. = 1.96 in (50 mm)
t = 0.042 in (1.07 mm)
L = 5.0 in (127 mm)

Figure 12.12 Tubes after loading in axial compression.

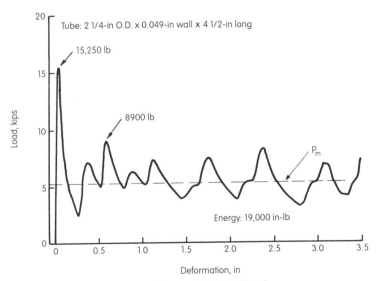

Figure 12.13 Static compression test of 6061-T6 tube.

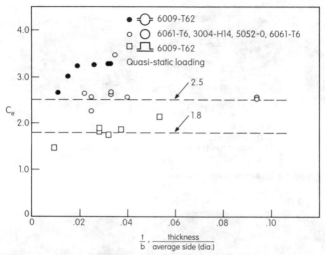

Figure 12.14 Effect of shape on energy absorption.

quent peaks are for other following folds. The area under the curve divided by the crushed distance is defined as the *mean load*.

Figure 12.14 gives coefficients and supporting data for the energy absorption of circular and rectangular shapes for use in Eq. (12.9). The results are from quasi-static tests. The energy absorbed in dynamic tests is higher than that for static tests, as shown by the coefficients in Fig. 12.15. Equation (12.9) is approximate; normally, tests of assemblies are conducted to verify final design.

12.3 Toughness/Ductility

"Toughness" and "ductility" have many interpretations, but usually these terms are considered to be related to the characteristics of the material. Some of the measured quantities that are considered are as follows:

- Fracture toughness
- Area under the stress-strain curve
- Elongation
- Reduction of area
- Unit propagation energy in a tear test
- Ratio of yield strength to tensile strength
- Charpy impact

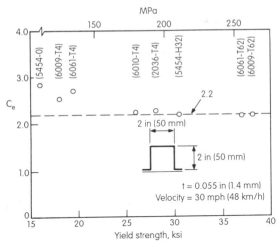

Figure 12.15 Effects of dynamic loading on energy absorption.

Only the first item is recognized as a characteristic that can be used by the structural designer of aluminum.[5] Fracture mechanics technology is useful for crack propagation studies to establish inspection intervals and to determine crack lengths for "brittle" failures. The high-strength aerospace alloys have requirements for fracture toughness. However, the general-purpose alloys considered here are too tough to be characterized by the use of fracture mechanics. Thus, the information available for the general-purpose alloys is one or more of the items on the above list, and there are no established ways to use the characteristic in design. All of the items have some value, however. Also, as pointed out previously, both material and design are important in a ductile or tough structure.

Area under the stress-strain curve and elongation are broad indicators of toughness (larger values are better) but are not very good for differentiating between alloys. Reduction of area seems to correlate somewhat with the ability of an alloy to accommodate a local strain (behavior of notched specimens in Chap. 5). Unit propagation energy correlates somewhat with fracture toughness values (also see Chap. 5). A low ratio of yield to tensile strength is beneficial to designers in that it allows the proportioning of joints that fail at a stress well above the yield strength of the member, providing substantial inelastic deformation and high energy to failure. Charpy impact tests have been commonly used for steels, particularly to establish transition temperatures. Aluminum structures have higher toughness (elongations) at low and high temperatures, compared to room temperature (Chap. 2), and, thus, have no transition temperature. Charpy values are therefore usually not available for the aluminum alloys.

At this time, the designer has few guides to alloy selection other than to use alloys that have performed well in similar applications. All of the general-purpose alloys discussed previously have good toughness and should perform well. Tests should be made for critical structures if the designer has any concern about structural toughness. It would be desirable to do more to incorporate toughness measures into practical design techniques.

12.4 Types of Corrosion; Corrosion Prevention

There are several types of corrosion attack that the designer needs to consider.[6] Considered below are (1) general weathering (uniform and pitting), (2) galvanic corrosion, (3) exfoliation, and (4) stress corrosion.

General weathering. Many aluminum products are left bare in the environment and perform satisfactorily. Examples are light poles, overhead sign structures, and small boats. These products are made from the alloys rated either A or B in Table 12.4, sometimes previously mentioned as general-purpose alloys. The most corrosive atmospheres are those in industrial areas and at the seacoast; rural areas are less severe. Corrosion attack in seacoast and industrial environments is self-limiting, but occurs at a faster rate at the seacoast. The oxides formed during weathering adhere to the base metal, and thus do not cause staining of adjacent structures. The bare structures of aluminum alloys acquire a light gray patina, normally quite satisfactory in ap-

TABLE 12.4   Relative Corrosion Ratings of Aluminum and Some of Its Alloys[6]

Alloy class	Commercial alloy example	Major alloying elements	General corrosion rating*
		Wrought, Strain-Hardened	
1XXX	1100	Unalloyed	A
3XXX	3003	Manganese	A
5XXX	5052, 5154	Magnesium	A
		Wrought, Heat-Treated	
6XXX	6061, 6063	Magnesium, silicon	B
2XXX	2027, 2017, 2024	Copper	D
7XXX	7075, 7178	Zinc, magnesium, copper	C

*Relative ratings are in decreasing order of merit. Alloys with B ratings can be used in industrial and seacoast atmosphere; alloys with lower ratings generally should be protected, especially on faying surfaces.

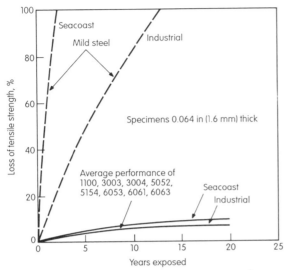

Figure 12.16   Loss of strength due to weathering.[6]

pearance. The color is somewhat different depending on the location and on the inclination of the surface.

The effects of the weathering are not sufficient to affect the structural integrity of aluminum structures. Figure 12.16 provides loss-of-strength data for thin specimens, exposed to corrosive environments to 20 years. A relatively small reduction in strength occurs for the aluminum alloys; less reduction would occur for thicker aluminum structures. Mild steel does not survive many years in these environments. The relatively small amount of attack experienced in test specimens and applications is obvious in Figs. 12.17 and 12.18 (some extrapolation of data to 52-year exposure).

There are a number of finishes that can be applied to aluminum if the appearance that results from natural weathering is not desired, including clear coatings, paints, and anodized coatings. For painting, proper surface preparation is needed. The structure to be painted must be cleaned with a solvent or other cleaner recommended for aluminum. Then the surface is primed with a chromate conversion coating or an acid etch wash primer. The final paint used must be compatible with the primer.

Galvanic corrosion (dissimilar materials).   Galvanic corrosion can occur when another metal such as steel is coupled to aluminum in the presence of an electrolyte. An aluminum part bolted to a steel structure with moisture allowed in the faying surface, or an aluminum part in concrete and coupled to the steel reinforcement are ex-

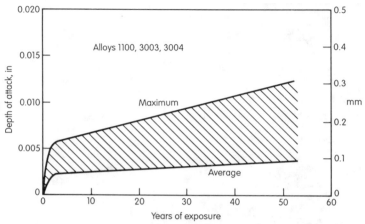

Figure 12.17   Attack at seacoast, tests and service.[6]

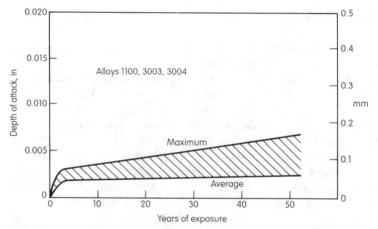

Figure 12.18   Attack at industrial site, tests and service.[6]

amples. The aluminum parts may act as an anode and be sacrificed in time. The attack can be prevented by isolating the two materials from each other. Practical steps are: (1) Prime and paint the steel contact surface using a zinc-based primer. (2) Keep the electrolyte out of the joint with a polysulfide, silicone, or butyl sealant. (3) Use galvanized or stainless steel (series 300) fasteners.

Aluminum in contact with porous materials that may absorb water and cause corrosion should be given a heavy coat of alkali-resistant bituminous paint or the equivalent. Contact with the earth is treated the same.

Exfoliation and stress corrosion.   The general-purpose alloys have excellent corrosion resistance and usually perform well. However, under heating over a period of time some of the 5XXX alloys may exhibit *exfoliation,* a delamination of the aluminum parallel to the surface. *Stress* corrosion, an attack resulting from tensile stress (can be a residual stress) and electrochemical reaction may also occur. 5XXX alloys containing over 3 percent magnesium (5083, 5086, and 5456) and subjected to elevated temperatures are susceptible to these two types of attack. Table 12.5[6] provides time and temperature information for alloys resistant to or susceptible to these types of corrosion. 5XXX alloys with less than 3 percent magnesium (5052, 5454) are resistant to both exfoliation and corrosion.

TABLE 12.5   Stress Corrosion and Exfoliation Resistance of Plate after Prolonged Heating[7]

Magnesium content, %	Alloy and temper	Heating time and temperature			Data available		Failures	
		Years	°F	(°C)	SCC*	Exf.†	SCC	Exf.
2.5	5052-F	12	150	(66)	Yes	Yes	None	None
		12	200	(93)	Yes	Yes	None	None
		12	250	(121)	Yes	Yes	None	None
2.7	5454-0	8	125	(52)	Yes	NA	None	NA‡
		8	150	(66)	Yes	NA	None	NA
		8	180	(82)	Yes	NA	None	NA
		8	200	(93)	Yes	NA	None	NA
		8	250	(121)	Yes	NA	None	NA
		8	300	(149)	Yes	NA	None	NA
	5454-H34	8	125	(52)	Yes	NA	None	NA
		8	150	(66)	Yes	NA	None	NA
		8	180	(82)	Yes	NA	None	NA
		8	200	(93)	Yes	NA	None	NA
		8	250	(121)	Yes	NA	None	NA
		8	300	(149)	Yes	NA	None	NA
3.5	5154-F	12	150	(66)	Yes	Yes	None	Yes
		12	200	(93)	Yes	Yes	None	Yes
		12	250	(121)	Yes	Yes	None	Yes
		12	300	(149)	Yes	Yes	None	Yes
	5154-0	2	250	(121)	Yes	NA	Yes	NA
	5154-H34	4	180	(82)	Yes	NA	Yes	NA
		4	200	(93)	Yes	NA	Yes	NA
		4	250	(121)	Yes	NA	Yes	NA
4.0	5086-0	4	150	(66)	Yes	NA	None	NA
4.5	5083-0	4	150	(66)	Yes	NA	None	NA
5.1	5456-0	4	150	(66)	Yes	NA	None	NA

*SCC based upon long transverse 0.125-in-diameter (3.2-mm-diameter) tensile bars stressed to 75% yield strength and exposed to 3½% NaCl solution by alternate immersion.
†Exfoliation (Exf.) based upon 24-h immersion exposure to solution of ammonium chloride, ammonium nitrate, ammonium tartrate, and hydrogen peroxide at 180°F (82°C).
‡NA indicates data not available.

The above is a brief review of some possibilities to prevent corrosion of aluminum structures. Much more information is available.[1,6–14]

References

1. *Alcoa Structural Handbook,* Aluminum Company of America, Pittsburgh, Pennsylvania, 1960.
2. Lengel, J. S., and Sharp, M. L., "Vibration and Damping of Aluminum Overhead Sign Structures," *Highway Research Record,* No. 259, 1969.
3. Thornton, C. P., "Wind Tunnel Tests of the Aerodynamically Induced Vibrations of Some Simple Structural Members,"[11] Mechanical Engineering Report MA-245, National Research Council of Canada, Ottawa, December 13, 1963.
4. Sharp, M. L., "Static and Dynamic Behavior of Welded Aluminum Beams," *Welding Journal Research Supplement,* February 1973.
5. *The Aluminum Association Position on Fracture Toughness Requirements and Quality Control Testing 1987,* The Aluminum Association, Washington, D.C., 1987.
6. Van Horn, Kent R. (ed.), *Aluminum,* Vol. 1, "Properties, Physical Metallurgy and Phase Diagrams," American Society for Metals, Metals Park, Ohio, 1967.
7. O'Shaughnessy, T. G., "Alcoa 5000-Series Alloys Suitable for Welded Structural Applications," Alcoa Green Letter No. 143, Aluminum Company of America, Pittsburgh, Pennsylvania, 1985.
8. McGeary, F. L., Englehart, E. T., and Ging, P. J., "Weathering of Aluminum," presented at the Northeast Regional Meeting of the National Association of Corrosion Engineers (NACE), Pittsburgh, Pennsylvania, October 6, 1965.
9. Lifka, B. W., "Corrosion Resistance of Aluminum Alloy Plate in Rural, Industrial and Seacoast Atmospheres," *Aluminum,* p. 1256, December 1987.
10. *Aluminum: The Corrosion Resistant Automotive Material,* T17, The Aluminum Association, Washington, D.C., 1986.
11. *Data on Aluminum Alloy Properties and Characteristics for Automotive Applications,* T9, The Aluminum Association, Washington, D.C., 1982.
12. *Repair and Maintenance of Aluminum Railcars,* No. 65, The Aluminum Association, Washington, D.C., 1984.
13. Hersh, J. F., "Corrosion Performance of Aluminum in Coal Railcars," paper presented at Corrosion 88, conference of the National Association of Corrosion Engineers, St. Louis, Missouri, March 21–25, 1988.
14. "Specifications for Aluminum Structures," *Aluminum Construction Manual,* Sec. 1, The Aluminum Association, Washington, D.C., 1986.

13

Design for Efficiency

The growth in the use of aluminum over the first 100 years of the industry has been primarily the result of replacement of incumbent materials not development of new products. This process of material selection is natural and will continue; for example, organic composites are currently challenging metals in some of the market areas. Replacement occurs because the new material provides a better solution to the problem (often economics) than the incumbent material. Whether the material is new or the incumbent, it is imperative that the designer develop the best design, often the most cost-effective design possible for the material.

This chapter contains information related to making efficient designs for aluminum products. Some of the comparisons are with steel, the major competing material to aluminum at this time. Some of the topics are (1) relative historical prices of aluminum and steel, (2) similarities and differences in the structural behavior of aluminum and steel, (3) designing for minimum weight, (4) designing for manufacturability, and (5) designing for life-cycle considerations.

13.1 Historical Prices of Aluminum and Steel

An analysis of the long-term substitution trends of several materials has been done.[1] For aluminum and steel, the price of pig iron and unalloyed aluminum ingot are used as the basis of comparison. Figures 13.1 and 13.2 show these comparisons based on equal weights and equal volumes, respectively. Most practical aluminum structures have weight savings of one-half to two-thirds compared to steel, falling in between equal weight and equal volume but generally much nearer to equal volume. Figure 13.2, thus, may be a better basis of comparison. The price of aluminum compared to steel has decreased

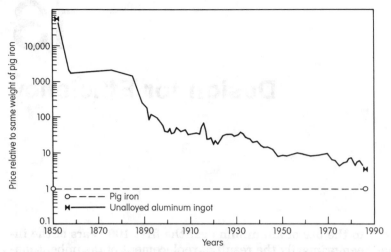

Figure 13.1 Relative price of unalloyed aluminum ingot and pig iron—same weight.[1]

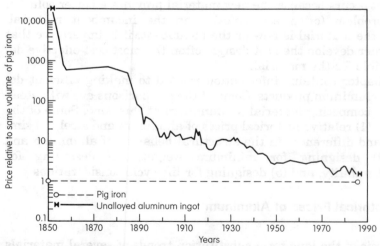

Figure 13.2 Relative price of unalloyed aluminum ingot and pig iron—same volume.[1]

significantly over the past 100 years. Fifty years ago, the price of aluminum was 5 to 6 times that of steel; it has decreased to 1 to 2 times that of steel in recent years. The significance of these changes is that as the price gap narrows, aluminum has become more attractive to the designer and customer in a new market area, and ultimately has replaced steel or the other incumbent material.

The general decrease in the price of aluminum relative to steel appears to be leveling off in recent years. Future prices will continue to depend on the improvements possible in the processes used to make the basic materials and to make the final product. The reduced price of the metal compared to steel in the twentieth century has been a significant factor in the growth of the use of aluminum.

13.2 Similarities and Differences in the Behavior of Aluminum and Steel

Both aluminum and steel are metallic, and thus many of their failure modes are similar. The same equations of elastic behavior apply to both materials. There are also some differences that the designer needs to be aware of. Table 13.1 presents a summary of characteristics of the two materials. The comments there apply to the normal

TABLE 13.1 Differences—Aluminum and Steel

Property	Steel	Aluminum	Importance for design
Modulus of elasticity	29×10^3 ksi $(200 \times 10^3$ MPa)	10.1×10^3 ksi $(70 \times 10^3$ MPa)	Deflection of members Vibration Buckling
Weight per volume	0.284 lb/in^3 $(0.078$ kg/cm^3)	0.10 lb/in^3 $(0.027$ kg/cm^3)	Weight of product Vibration
Thermal expansion	7×10^{-6} in/in/°F (12.6×10^{-6})	13×10^{-6} in/in/°F (23.4×10^{-6})	Thermal expansion Thermal stress
Stress-strain curves	Varies	Varies	Depends on alloys Steel often has higher strength and elongation at room temperature Aluminum has better performance at low temperatures
Fatigue	Varies	Varies	For joints, aluminum has about ⅓ to ½ fatigue strength as steel for same detail
Corrosion resistance	Needs protection	Often used unpainted	Aluminum usually is maintenance-free Aluminum is nonstaining
Strain rate effects— mechanical properties	High strain rates increase properties—varies with type of steel	Much less change in properties compared to steel	Need to use dynamic properties for high-strain-rate loadings

$172 \frac{\#}{Cu\ FT}$

general-purpose materials used in many applications. Exceptions can be found to most of the comments if the entire range of steels and aluminum alloys are included.

The modulus of elasticity of aluminum is about one-third that of steel. The primary areas of structural design that are affected are deflection of members, buckling, and vibration. The mass of aluminum is also about one-third that of steel, so that the weight of aluminum parts is much less than those of steel. Because the ratio of mass to modulus is about the same for aluminum and steel, the natural frequencies of aluminum and steel parts of identical geometry are essentially the same.

Thermal expansion of aluminum parts is about twice that of steel, so more allowance for movement due to temperature change needs to be considered for expansion devices for aluminum structures, for example. If thermal expansion is restrained, the stress caused by a given temperature change will be less for aluminum than for steel because of the lower modulus of elasticity of aluminum. Stress-strain curves for general-purpose steels and aluminum alloys show that steel often has higher tensile strength and elongation at room and elevated temperature, but aluminum is much better in ductile performance at low temperatures.

Fatigue strengths of aluminum joints are about one-third to one-half those of identical geometries in steel. Aluminum parts have a very durable oxide surface and excellent corrosion resistance either bare or coated. Steels need protection in most environments; otherwise, severe rusting occurs. High strain rates increase the mechanical properties of steel structures more than they increase those of aluminum.

An understanding of the differences is important because the designs of products of aluminum and steel must be different in order to produce cost-effective solutions for either material. Designing for efficiency is covered in the subsequent sections.

13.3 Designing for Minimum Weight

In the design of aluminum structures for weight efficiency, a general rule of thumb is that the sections should be larger than those of steel. When deflection and fatigue are critical, the weight savings with aluminum should be about 50 percent. Figure 13.3 presents a simple analysis of the weight and size of thin-walled box sections that provides some verification for the above rules. The two curves are based on equal moment capacities of aluminum and steel sections and equal deflections of the two types of sections. The yield strengths and loads are assumed to be equal. The equations employed are as follows:

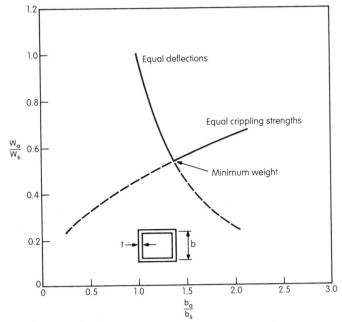

Figure 13.3 Minimum weight of square tubular sections.

For moment causing crippling failure:

$$M_a = \sigma_a S_a \qquad M_s = \sigma_s S_s \qquad\qquad (13.1)$$

where $\sigma_a \propto \sqrt{E_a \sigma_{ya}}(t_a \,/\, b_a)$

$\sigma_s \propto \sqrt{E_s \sigma_{ys}}(t_s \,/\, b_s)$

subscript a = aluminum
subscript s = steel
M_a, M_s = moment at failure
σ_a, σ_s = crippling stress
σ_{ya}, σ_{ys} = yield strength
S_a, S_s = section modulus
b_a, b_s = length of side of square tube
t_a, t_s = thickness of square tube
E_a, E_s = modulus of elasticity

For equal moments: $\dfrac{W_a}{W_s} = \dfrac{\rho_a}{\rho_s}\left(\dfrac{b_a}{b_s}\right)^{1/2}\left(\dfrac{\sigma_{ys}}{\sigma_{ya}}\right)^{1/4}\left(\dfrac{E_s}{E_a}\right)^{1/4}$ (13.2) where W_a, W_s = weight

ρ_a, ρ_s = unit weight

For deflection of member: $\delta_a \propto \dfrac{V_a}{E_a I_a}$ $\delta_s \propto \dfrac{V_s}{E_s I_s}$ (13.3)

where δ_a, δ_s = deflection

V_a, V_s = load

$I_a \propto b_a^3 t_a$, $I_s \propto b_s^3 t_s$ = moment of inertia

For equal deflection: $\dfrac{W_a}{W_s} = \dfrac{\rho_a}{\rho_s} \dfrac{V_a}{V_s} \dfrac{E_s}{E_a} \left(\dfrac{b_s}{b_a}\right)^2$ (13.4)

Crippling strength is used so that abnormally thin sections are not allowed as solutions. The same crippling formula is employed for the two materials for simplicity although it is approximate for steel sections. Figure 13.3 shows that minimum weight occurs in this case when the aluminum section is about 40 percent larger than the steel section. The weight savings for the aluminum part is 46 percent. If the load on the aluminum section is less than that on the steel section, the size of the aluminum section is smaller and the weight savings more than that shown in Fig. 13.3. Note that there is no weight savings if the sections are the same size, because the aluminum section must be much thicker than the steel section to offset the modulus difference.

Another type of structure to consider is one in which the self-weight is the primary load carried. Long-span bridges are practical examples of this type of structure. Equations have already been developed for this case.[2] The equations presented below are based on this work. Failure was considered to be caused by buckling of the girder web. There was no limitation on deflection.

For long-span girders: other loads are negligible compared to the weight of the girder[2]

$$\dfrac{W_a}{W_s} = \left(\dfrac{\rho_a}{\rho_s}\right)^3 \left(\dfrac{E_s}{E_a}\right)^{1/2} \left(\dfrac{\sigma_{ys}}{\sigma_{ya}}\right)^{3/2}$$ (13.5)

For short-span girders: weight of girder is negligible compared to the other loads[2]

$$\dfrac{W_a}{W_s} = \dfrac{\rho_a}{\rho_b} \left(\dfrac{E_s}{E_a}\right)^{1/6} \left(\dfrac{\sigma_{ys}}{\sigma_{ya}}\right)^{1/2}$$ (13.6)

Figure 13.4 presents the results using Eqs. (13.5) and (13.6). The girders are optimally proportioned for aluminum and steel, subject to the assumptions. Allowable stresses appropriate to bridge structures are used in Fig. 13.4 rather than yield strengths. The short-span beams

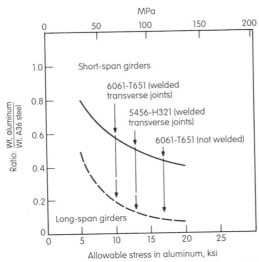

Figure 13.4 Relative weights of aluminum and A36 steel girders of equal strength. [*Note:* Allowable stress of 20 ksi (138 MPa) used for A36 steel.]

have a weight of about 50 percent that of steel (comparable strengths of materials), consistent with the previous study. Long-span aluminum structures have the potential to weigh much less than those of steel (less than 10 percent indicated on Fig. 13.4), depending on the strengths of the materials. Obviously the lower unit weight of aluminum will affect the design in those cases in which the self-weight is significant compared to the imposed loads.

Automotive body panels such as hood and decklid assemblies have requirements for bending and torsional stiffness, dent resistance, vibration resistance, and resistance to permanent set. In designing aluminum auto body panels, rather than increasing thickness of the incumbent steel design to meet the requirements, it is more weight efficient to change the design, particularly the inner-panel geometry.[3] Figure 13.5 shows one case in which the outer-panel thickness is reduced to the minimum possible value at which it can meet dent-resistance requirements. The inner panel is changed correspondingly, with more subribs and increased flexural stiffness of the ribs to meet the remainder of the requirements. Table 13.2 provides a summary of the methods to minimize weights of auto body panels.

Another way to design for minimum weight is to proportion the cross section of the shape in the most efficient manner to carry the load. One example is provided in Fig. 13.6. Shown are contours of the ratio of the buckling strength of a simply supported plate with vary-

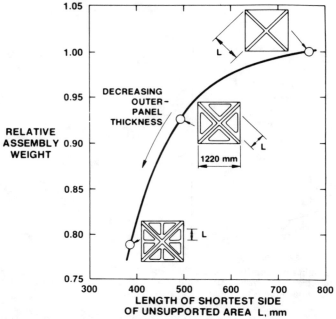

Figure 13.5 Effect of inner-panel rib spacing on relative weight of auto body hood or decklid assembly. (*Reprinted with permission from SAE Paper No. 790164, © 1979 Society of Automotive Engineers, Inc.*)

TABLE 13.2 Guidelines for Minimizing the Weight of Aluminum Auto Body Panels

Performance criterion	Guideline for minimum weight
Denting	Increase outer-panel yield strength.
Local stiffness (oil-canning)	Reduce inner-panel rib spacing.
Torsion and bending stiffness	Increase inner-panel rib depth. Select most desirable rib orientation.
Resistance to permanent set	Increase inner- and outer-panel yield strength. Increase inner-panel rib depth.
Crippling	Increase inner-panel yield strength. Select most desirable inner-panel rib geometry.
Resistance to forced vibration of outer panel	Reduce inner-panel rib spacing.
Resistance to forced vibration of assembly	Increase inner-panel rib depth.

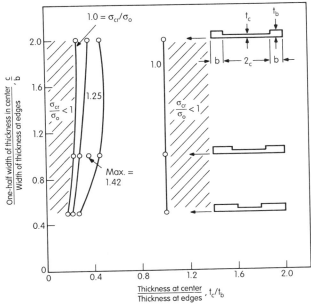

Figure 13.6 Effect of varying thickness of plate on compressive buckling. σ_{cr} = buckling strength of simply supported plate with varying thickness; σ_o = buckling strength of simply supported plate of uniform thickness with same cross-sectional area as corresponding plate of varying thickness.[4]

ing thickness to that for a plate with uniform thickness, both having the same area.[4] The plates are subject to edge compression; the sketches in Fig. 13.6 show cross sections normal to the direction of applied loading. The plate with varying thickness of the proper geometry has a buckling load 42 percent higher than that for the plate with uniform thickness. Thus, plates thicker at the edges than at the center are much more efficient in compressive buckling than those with uniform walls. These designs are easy and economical to achieve with aluminum extrusions. They would be difficult to manufacture in steel.

Figure 13.7 gives a few more possibilities for minimizing weight. Normally, designers think in terms of the constant-thickness walls that many standard products have, but with extrusions, stiffening lips can be added to stabilize thin flanges, and intermediate stiffeners can be used to improve buckling efficiency.

There are many ways to minimize the weight of aluminum parts. Minimum weight is usually cost effective for most applications although the cost of manufacturing is also important.

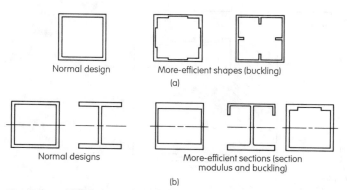

Normal design   More-efficient shapes (buckling)

(a)

Normal designs   More-efficient sections (section
                        modulus and buckling)

(b)

Figure 13.7 Efficient sections for aluminum. (*a*) Shapes in compression, (*b*) shapes in bending.

13.4 Designing for Manufacturing

Three examples are provided to illustrate the need to consider cost-effective manufacturing in the early design phase. Figure 13.8 shows a typical side roof rail of an automobile. In steel it would consist of three stamped pieces spot-welded together to make the tubular assembly. The aluminum extrusion shown on the right takes the place of three parts. The economics are good because the cost of the extrusion die is low compared to the stamping tools. One-third as many parts are handled in the assembly process. The three rows of spot welds needed to join the three sheet stampings are eliminated. Finally, the torsional and flexural stiffnesses are better in the extrusion because the joints in sheet construction have flexibility on account of the discrete attachments, resulting in a more flexible member.

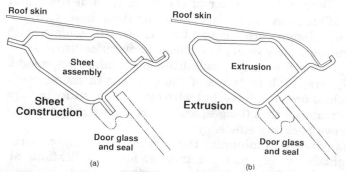

Roof skin   Roof skin

Sheet
assembly   Extrusion

Sheet
Construction   Extrusion

Door glass
and seal   Door glass
                  and seal

(a)   (b)

Figure 13.8 Alternate construction for automotive roof rail. (*a*) Typical side roof rail, (*b*) modified side roof rail.

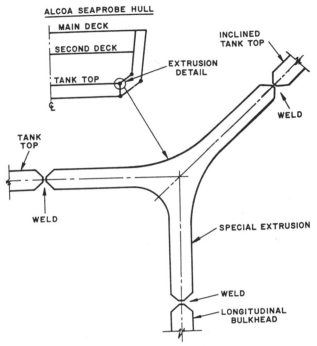

Figure 13.9 Special extrusion to attach plating in Alcoa *Seaprobe*.

Figure 13.9 shows an extrusion used in the construction of the Alcoa *Seaprobe*. The member is used in the bottom structure of the vessel at a location in which three plates intersect, a detail that causes difficulty in achieving good-quality welds. The extrusion allows simpler joints to be made, away from the intersection. The extrusion is shaped at the attachment points to make the welds. The fatigue performance is improved because better welds are placed in a less stressed region.

An extrusion used as part of the deck on the Smithfield Street Bridge in Pittsburgh is shown in Fig. 13.10. A number of features intended to facilitate construction and erection are designed into the extrusion. A clearance at the bottom of the section was incorporated to accommodate rivet heads in the existing floor beams. The spacing of the bolting flanges and the depth of the clearance were selected based on the size of the rivets and their spacing. The upper parts of the extrusion that are attached to the plate were shaped to facilitate welding. From the standpoint of performance, the closed stiffener, when joined to the plate, helped to transfer the effects of the wheel loads lat-

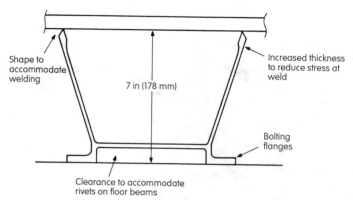

Figure 13.10   Extrusion for Smithfield Street Bridge deck.

erally over several ribs. The localized thickening at the upper leg reduced the stresses in the weld, thereby improving fatigue strength.

13.5   Life-Cycle Considerations

Some discussion of life-cycle cost was included in Chap. 3. Cost is one of the important considerations in design, but there are other issues that are more important now than ever. One issue that is getting worldwide attention is the detrimental effects that air pollution is having on our environment. What were once scientific curiosities about the possibility of global warming and depletion of the earth's ozone layer, for example, now are concerns shared by many scientists about how much the quality of future life might be affected even if society starts to correct some of our current practices. Future designers of products will need to consider the long-term consequences of their design.

One of many products that contribute to pollution of the air is the automobile. Studies have been published[5] to quantify lifetime energy consumption for aluminum and steel parts on the vehicle. Figure 13.11 shows that the energy to fabricate the parts is less for steel than aluminum. However, the lifetime use of energy will be much less for aluminum parts. The lighter weight means less fuel is needed to move the vehicle during its lifetime, and the largest portion of the total energy use is to operate the vehicle, not to build it. Reducing the amount of fuel used also reduces air pollution.

The life-cycle considerations can now be listed with a broader charge for the designer: to search for the cost-effective solution, but one that does not create or sustain problems for society now or in the future.

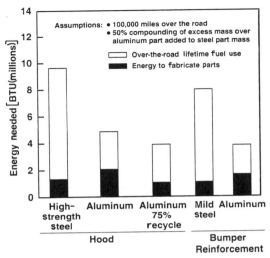

Figure 13.11 Lifetime energy consumption.[5]

Life-cycle considerations include

- Manufacture of the basic materials for the product
- Fabrication of the materials into a useful configuration
- Assembly or erection of the components into the final product
- Maintenance and/or operation of the product over its useful life
- Disposal of the product after its useful life

13.6 Potential for Use of Aluminum in Long-Span Bridges—Case Study

High strength-to-weight ratio and maintenance-free construction are factors that may favor aluminum in this application. In addition, aluminum maintains high strength and toughness at low temperatures. On the other side, the low modulus of elasticity, high thermal expansion, and high cost per unit weight of aluminum compared to steel are possible disadvantages. No long-span aluminum bridges exist, although several short-span bridges of various types of construction have been in service for around 40 years.[6] This brief, preliminary study was originally done a number of years ago but is included here because it still gives some valid input on the potential for use of aluminum in this application.

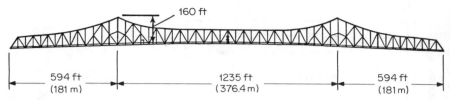

Figure 13.12 Elevation: cantilever truss bridge.

Figure 13.12 shows the elevation of the steel bridge redesigned in aluminum. The width of the bridge is 94 ft (28.7 m), center-to-center of the trusses. The weights of the various parts in steel are given in Table 13.3. Approximately 80 percent of the load in highly stressed members is from the dead loads. The live-load deflection is about ⅟₁₃₀₀ of the main span. Stringers and floor beams were redesigned in aluminum using plate fabrications of alloy 5456-H321. Figure 13.13 shows a representative design. The weight of the aluminum shapes is about 50 percent that of steel.

The aluminum trusses were designed by trial. The reduction of dead weight due to the light weight of the aluminum members was taken into account. A representative member in 6061-T6 is given in Fig. 13.14. No attempt was made to optimize the geometry of the aluminum design, although some improvement in design that decreases the amount of material should be possible. The weights of the aluminum parts are given in Table 13.4. The weight of the aluminum structure is

TABLE 13.3 Summary of Dead Loads on Steel Truss Bridge

Weight of all steel, lb (kg)		30,072,500 (13,669,300)
Weight of concrete deck, lb (kg)		22,372,200 (10,169,200)
	Total dead weight, lb (kg)	52,444,700 (23,838,500)
Trusses (59%)		
HS50	1,145,000 (520,500)	
HS60	9,038,000 (4,108,200)	
A440	2,975,000 (1,352,300)	
A36	4,614,430 (2,097,500)	17,772,430 (8,078,400)
Floor beams and stringers (25%)		
A36	7,400,070 (3,363,700)	7,400,070 (3,363,700)
Lateral framing (16%)		
A441	900,000 (409,000)	
A36	4,000,000 (1,818,000)	4,900,000 (2,227,000)
	Weight of all steel	30,072,500 (13,669,300)

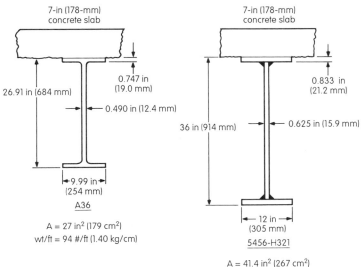

Figure 13.13 **Stringer design.**

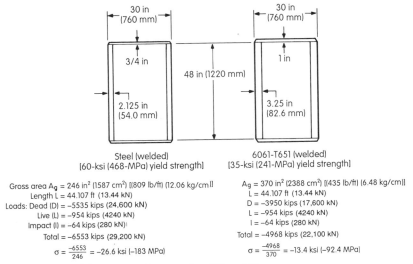

Figure 13.14 **Main compression member ($L/r = 40$).**

TABLE 13.4 Weights of Aluminum and Steel in Bridge of Same Geometry

	Steel, lb (kg)	Aluminum, lb (kg)	
		6061-T651*	5456-H321†
HS50	1,145,000 (520,500)	526,700 (239,400)	
HS60	9,038,000 (4,108,200)	4,699,760 (2,136,300)	
A440	2,975,000 (1,352,300)	1,368,500 (622,000)	
A36	16,014,510 (7,279,300)	1,707,340 (776,100)	5,700,040 (2,590,900)
A441	900,000 (409,000)		450,000 (204,500)
	30,072,500 (13,669,300)	8,302,300 (3,773,800)	6,150,040 (2,795,500)
		14,452,340 (6,569,300)	

(Weight of aluminum)/(weight of steel) = 0.48. Deflection of aluminum truss ≅ 1/640 span.
*Main trusses.
†Floor beams, stringers, cross bracing.

about 48 percent that of steel. The load in highly stressed aluminum members is about 75 percent that in steel, reflecting the lighter weight of the aluminum framing. Higher-strength aluminum could be considered to reduce weight further, but the live-load deflections would also have been larger. Other design changes such as replacing the concrete deck with a much lighter weight aluminum system could also be made, but the purpose of this study was to compare similar systems. The design in aluminum would meet the allowable stress requirements for maximum loads or for fatigue as defined in the AASHTO guide specifications. Thermal expansion was assumed to be accommodated by proper design of expansion devices.

The cost of any new application is difficult to estimate. Table 13.5 presents some relative values. The values for steel for the different items are believed to be reasonable for a preliminary assessment. Some of the reasons for the values assumed for aluminum follow. The

TABLE 13.5 Cost of Bridges

Item	Steel	Aluminum
Metal	$W_s C_s$	$W_a C_a$
Fabrication	$1.92\, W_s C_s$	$1.92\, W_s C_s$
Erection	$W_s C_s$	$0.90\, W_s C_s$
Cleaning or painting	$0.025\, W_s C_s$	$0.03\, W_s C_s$
Total	$4.17\, W_s C_s$	$W_a C_a + 2.85 W_s C_s$

W_s, W_a = weights of steel and aluminum, respectively; C_s, C_a = metal cost per pound of steel and aluminum, respectively.

first item is simply the cost of the base metal times the total weight of the metal in the bridge.

The cost of fabrication of the aluminum and steel structures was assumed to be the same. There are trade-offs in making this assumption. More care in cleaning is required for aluminum than steel in welding, and the welds are larger in aluminum. Because flame cutting is not allowed for aluminum, cutting of thick sections may be more costly for aluminum. On the other hand, aluminum can be saw-cut and drilled readily, with less wear on tools than for steel. Because of the percent smaller load in the aluminum members, fewer bolts and holes may be needed. Because the weight of aluminum parts is about half that of steel, some savings in handling may also be possible.

The lightness of aluminum structures compared to steel is a decided advantage in field erection. There is the probability of prefabrication of larger portions of the bridge in aluminum. The lower loads mean fewer bolts are needed in the field erection. A modest 10 percent savings in erection costs was assumed for the aluminum bridge.

The steel bridge is painted. The cost can vary appreciably depending on the type of steel and type of protection. The aluminum is cleaned at a much smaller cost.

From Table 13.5 the relative cost of the erected steel and aluminum structures is calculated as follows:

$$\frac{\text{Total aluminum structure cost}}{\text{Total steel structure cost}} = 0.24 \frac{W_a C_a}{W_s C_s} + 0.68 \qquad (13.7)$$

The weight of aluminum compared to that of steel was 0.48, as previously mentioned. For a C_a/C_s of 4, Eq. (13.7) shows that the first cost of the aluminum structure is about 1.14 times that of steel. Because the estimates do not include the future cost of maintenance, possible lower costs of foundations, and so forth, the cost is close enough that more detailed studies should be made, particularly for structures more favorable to aluminum. Cases of interest would be longer spans than that considered here, in which the light weight of aluminum would allow more reduction of weight as compared to steel.

References

1. Cochran, C. N., "Long-Term Substitution Dynamics of Basic Materials in Manufacture," *Materials and Society*, Vol. 12, No. 2, pp. 125–150, 1988.
2. Haaijer, G., "Economy of High Strength Steel Structural Members," Paper No. 3509, *Transactions ASCE*, Vol. 128, Part II, 1963.
3. Rolf, R. L., Sharp, M. L., and Herbein W. C., "Minimizing the Weight of Aluminum Body Panels," SAE Paper 790164, Society of Automotive Engineers, Warrendale, Pennsylvania, 1979.
4. Capey, E. C., "The Buckling under Longitudinal Compression of a Simply Supported

Panel that Changes in Thickness across the Width," Royal Aircraft Establishment, Technical Note No. Structures 174, Ministry of Aviation, London, England, August 1956.
5. *Use of Aluminum in Automobiles—Effect on the Energy Dilemma*, T12, The Aluminum Association, Washington, D.C., 1980.
6. Clark, J. W., "Aluminum Brides—An Evaluation," Meeting preprint MTL-58, ASCE/EIC/RTAC Joint Transportation Engineering Meeting, Montreal, July 15–19, 1974.

14

Codes, Standards, and Specifications for Aluminum Structures

One of the references most important to the designer—essential for many applications—is a code or standard that provides guidance for good practice. These documents are generally developed with good input from practicing engineers familiar with the historical problems for the type of structure considered. After a period of time, they reflect the experience of a number of designers and the practice leading to satisfactory performance of the product. In most cases, codes and specifications contain mandatory provisions; the design must be done using the methods provided.

The purpose of this chapter is to list codes, specifications, and other documents that designers should consider as references, and to provide sufficient information that the designer can make a judgment as to whether or not a particular reference is useful for his or her work. Information on where one can obtain a copy of each work is also given.

14.1 Aluminum Association Specifications

There are four documents published by The Aluminum Association pertinent to the design of aluminum structures. They will be discussed here.

Specifications for Aluminum Structures (Construction Manual Series, Sec. 1, No. 30, 1986; The Aluminum Association Inc., 900 19th Street N.W., Washington, D.C. 20006)

The specification was originally developed to apply to building and highway structures with factors of safety essentially the same as those in corresponding specifications for steel. It has the allowable stress format and is probably the most used document in the United States for aluminum structures. The specification covers all the suggested general-purpose alloys and guidelines for designing tension members, columns, components of columns, beams, components of beams, mechanical fastening, and welded construction. Minimum mechanical properties suitable for design use are included. National and local building codes and highway product design codes reference The Aluminum Association's specifications for structural design information.

Engineering Data for Aluminum Structures (Construction Manual Series, Sec. 3, No. 33, 1986; The Aluminum Association Inc., 900 19th Street N.W., Washington, D.C. 20006)

This publication contains engineering data that supplements the information provided in the specification. Included are typical mechanical and physical data for most of the alloys of general structural use, including bend radii for sheet and plate. Tables of properties of wide-flange sections, I-beams, channels, angles, tees, zees, round and square tubes, and some special shapes are provided. Tables of allowable loads and spans for I-beams, treadplate, and commercial roofing and siding products are included. Formulas for bending and deflection of beams under various loading and support conditions are given. Also provided are types of rivets, installation information, bolt and nut dimensions, and strength information on fasteners.

Illustrative Examples of Design (Construction Manual Series, Sec. 2, No. 32, 1990; The Aluminum Association Inc., 900 19th Street, N.W., Washington, D.C. 20006)

This publication presents detailed design calculations for a number of practical aluminum structures: tension members, columns, girders, formed sheet siding, etc. The examples are selected to illustrate the major provisions of The Aluminum Association "Specifications for Aluminum Structures." The design examples also make use of the information given in the "Engineering Data for Aluminum Structures."

Commentary on the Aluminum Association Specifications (Construction Manual Series, Sec. 1A, No. 31, The Aluminum Association Inc., 900 19th Street N.W., Washington, D.C. 20006, 1982)

This document contains the background information supporting each of the provisions covered in the specifications. Also included is a comprehensive list of references pertinent to the development of the specifications.

14.2 Pressure Vessels and Piping

Section VIII, Rules for Construction of Pressure Vessels (Division 1, July 1, 1989; The American Society of Mechanical Engineers, United Engineering Center, 345 East 47th Street, New York, NY 10017)

This document covers the design of pressure vessels, containers subject to either internal or external pressure. Subsection A gives general requirements applicable to all materials and construction, such as equations for determining stresses in the body and heads of the container, details for openings, permissible out-of-roundness, inspection, testing, and reports. Subsection B of the document gives design information for fabrication of vessels by welding, forging, and brazing. Subsection C gives requirements according to classes of materials. Aluminum is covered in Part UNF (normal use) and Part ULT (low-temperature use) of Subsection C.

Also included in Subsection C are tables of allowable tensile stresses, depending on temperature. Several appendices provide mandatory and nonmandatory provisions. Included are design formulas, design charts for thickness of parts, required details, inspection standards, and design examples.

ASME B31.3b-1991 Addenda to ASME B31.3 (1990 edition, Chemical Plant and Petroleum Refinery Piping, The American Society of Mechanical Engineers, 345 East 47th Street, New York, NY 10017, 1990)

This code for pressure piping applies to piping for all fluids and gives requirements for materials, design, fabrication, assembly, erection, examination, inspection, and testing. Equations for stress calculation, including fatigue, and allowable design stresses are included.

14.3   Storage Tanks

API Standard 620, Recommended Rules for Design and Construction of Large, Welded, Low-Pressure Storage Tanks (Revision 1, April 1985; American Petroleum Institute, 1220 L Street N.W., Washington, D.C. 20005)

This document provides rules on materials, design, fabrication, inspection and testing, and pressure- and vacuum-relieving devices. Aluminum alloys are referenced in Appendix Q concerning storage tanks for liquefied gases. Design methods, good practices for details, and properties are provided to the designer.

ASME/ANSI B96.1-1986, Welded Aluminum-Alloy Storage Tanks (1986; The American Society of Mechanical Engineers, United Engineering Center, 345 East 47th Street, New York, NY 10017)

The Standard covers the design, materials, fabrication, erection, inspection, and testing requirements for welded aluminum alloy, field-erected or shop-fabricated, aboveground, vertical, cylindrical, flat-bottom, open- or closed-top tanks storing liquids under pressures approximating atmospheric pressure at ambient temperatures. Equations for stress calculation, allowable stresses, and good practices for details are included.

14.4 Ship Structures

Rules for Building and Classing Aluminum Vessels (1975; American Bureau of Shipping, 45 Broad Street, New York, NY 10004)

The rules apply to vessels 100 ft (30.3 m) to 500 ft (152.5 m) long built of aluminum alloys. The rules give specific requirements for thickness and size of all major structural parts of a ship. Suitable aluminum alloys and their tensile properties are listed. Information is also provided on various aluminum product forms, welded construction, and corrosion considerations.

14.5 Highway Structures

Standard Specifications for Structural Supports for Highway Signs, Luminaires and Traffic Signals (1985; American Association of State Highway and Transportation Officials, 444 North Capital Street N.W., Suite 225, Washington, D.C. 20006)

This specification gives wind and ice loads for various structures, methods of analysis, aluminum design (Sec. 5), and guidance on breakaway supports, foundations, and other details of design. Several shapes of members are included in the considerations. The specification covers the supporting structure only (not signs, etc.).

Guide Specifications for Aluminum Highway Bridges (1991; American Association of State Highway and Transportation Officials, 444 North Capital Street N.W., Suite 225, Washington, D.C. 20006)

This guide specification was patterned after the AASHTO specification for steel, in format and in the breath of coverage of the provisions. A limited number of alloys were included in the allowable stress tables, but there is much detail in the design provisions for proportioning components of bridge structures. Sections on fabrication of aluminum structures are also included.

14.6 Aircraft Structures

MIL HDBK 5E (Military Standardization Handbook, Metallic Materials and Elements for Aerospace Vehicle Structures, Vol. 1, Chap. 3, Naval Publications and Forms Center, 5801 Tabor Avenue, Philadelphia, PA 19111-5094)

This publication contains design allowable values for aluminum alloys of interest in aerospace applications. Included are mechanical properties, such as tensile and compressive stress-strain behavior, modulus of elasticity, fatigue of notched and smooth specimens, crack propagation, etc. Effects of environment and temperature are included in some cases. This is the most comprehensive listing of design properties for aerospace alloys in existence in the United States.

14.7 General Design Information

Structural Design with Aluminum (No. 38, 1987; The Aluminum Association, 900 19th Street N.W., Washington, D.C. 20006)

This booklet provides an introduction to the design of aluminum structures, with some discussion concerning the similarities and differences between aluminum and steel. Included are properties of a few common alloys, design examples, joining methods, and other design considerations such as corrosion.

Aluminum Standards and Data (No. 1, 1990; The Aluminum Association, 900 19th Street N.W., Washington, D.C. 20006)

This publication provides minimum and typical tensile properties for all commercial alloys and product forms. Data are also given for chemical composition, heat-treatment and aging practices, and standard tolerances of various mill products.

14.8 Fabrication Processes

ANSI/AWS D3.7-83, Guide for Aluminum Hull Welding (1983; American Welding Society, Inc., 550 N.W. LeJeune Road, Miami, FL 33126)

The main sections in this publication cover welded aluminum hulls, aluminum hull materials, preparation for construction, welding processes and equipment, qualification procedures for welding, and welding procedure and techniques.

ANSI/AWS D1.2-90, Structural Welding Code, Aluminum (1990; American Welding Society, 550 N.W. LeJeune Road, P.O. Box 351040, Miami, FL 33135)

The code contains sections on general provisions, design of welded connections, workmanship, technique, qualification of procedures and personnel, inspection, stud welding, nontubular statically loaded structures, nontubular dynamically loaded structures, strengthening and repair of existing structures, appendices, and commentary.

14.9 Canadian Standards

Ontario Highway Bridge Design Code (1983; Ministry of Transportation and Communications, 1201 Wilson Avenue, Downsview, Ontario, Canada M3M 1J8)

The document gives general requirements for bridges and other transportation-related structures. Aluminum is specifically mentioned for highway supports. Fatigue curves are given for several types of aluminum joints. CSA Standard S157-M is referenced for other design information on aluminum.

Strength Design in Aluminum (1983; National Standard of Canada, CAN3-S157-M83, Canadian Standards Association, 178 Rexdale Boulevard, Rexdale (Toronto), Ontario, Canada, M9W 1R3)

Covers mechanical properties of aluminum alloys and design methods for tension and compression, bending, and combined loading members. Effects of joining on member behavior are included. The breadth of treatment is similar to that in The Aluminum Association "Specifications for Aluminum Structures" and related documents. Also, the Canadian Standard is presented in the limit state format.

Glossary

A Area of component

A_F Area of flange

A_r Area of reduced-strength material

A_s Area of stiffener

a One-half of crack length; or smaller dimension of shear panel; or width of web corrugation; or length of side of square

B_b Buckling formula constant for bending of rectangular bar

B_c Buckling formula constant for columns

B_p Buckling formula constant for plates

B_s Buckling formula constant for tubes and plates in shear

B_t Buckling formula constant for tubes in compression

B_{tb} Buckling coefficient for tubes in bending

b Stiffener spacing, flange width, larger dimension of shear panel; or width of plate

b_a Width of aluminum section

b_s Width of steel section

b_w Width of web

C Buckling coefficient for plates; or centroid; or buckling coefficient for corrugated webs; or coefficient for buckling of cylinders; or coefficient for pull-through strength

C_1 Coefficient for girder design; or constant in web-crippling equation; or coefficient for shear deformation of corrugated panels

C_{1e} Constant in web-crippling equation

C_2 Coefficient for girder design; or constant in web-crippling equation; or coefficient for shear deformation of corrugated panels

C_a Cost of aluminum per unit weight; or stress added to buckling stress for appearance

C_b Buckling formula constant for rectangular sections in bending

C_c Buckling formula constant for columns

C_e Coefficient for energy absorption

C_m Coefficient for cases of nonuniform bending

C_s Cost of steel per unit weight

C_v Coefficient for frequency

C_w Warping constant

c Length of lip

D Plate-bending stiffness; or dead load; or diameter; or diameter of washer; or maximum static deflection

D_b Buckling formula constant for rectangular sections in bending

D_c Buckling formula constant for columns

D_p Buckling formula constant for plates

D_s Sheet-bending stiffness; or buckling formula constant for tubes and plates in shear

D_t Buckling formula constant for tubes in compression

D_{tb} Buckling formula constant for tubes in bending

D_w Plate-bending stiffness of web

D_x Plate-bending stiffness in x direction

D_y Plate-bending stiffness in y direction

d Diameter of fastener, diameter of hole; or diameter; or depth of section

E Modulus of elasticity

e Edge distance

F Factored load

F_s Compressive force in stiffeners

f Frequency

f_b Calculated flexural stress

G Modulus of rigidity; or shear modulus

G' Shear rigidity of assembly

g Acceleration of gravity

h Distance to centroid of combined lip and flange; or web depth

I_e Moment of inertia of stiffener plus width of plate equal to the stiffener spacing

I_p Polar moment of inertia

I_s Moment of inertia of stiffener

I_x Moment of inertia about x axis

I_{xo} Moment of inertia about xo axis; or moment of inertia about centroidal axis

I_y Moment of inertia about y axis

I_{yc} Moment of inertia of flange and lip about centroidal axis; or moment of inertia of compression flange

I_{yo} Moment of inertia about *yo* axis; or moment of inertia about centroidal axis

J Torsion constant

K Buckling coefficient for orthotropic plate; or coefficient for bursting of pipe; or effective-length coefficient for columns

K_1 Constant that defines slenderness ratio transition between buckling and crippling; or factor for frequency of vibration

K_2 Constant for crippling strength

K_{1c} Plane-strain fracture toughness

K_c Nonplane-strain fracture toughness

K_ϕ Effective-length coefficient for torsional buckling; or elastic restraint factor

k Coefficient for plate buckling

k_b Coefficient for lateral buckling

L Length of column; or unsupported length of beam; or length of panel; or live load; or length

L_c Length in direction of corrugation

M Moment

M_1 Applied ultimate moment; or moment at end 1

M_2 Moment at end 2

M_a Moment at failure for aluminum part

M_c Critical moment

M_s Moment at failure for steel part

M_u Ultimate moment

N Number of panels in stiffened plate; or length of load

NS Notch strength

N_g Fatigue life under spectrum loads

N_i Fatigue life of stress level, σ_i

P Load; or shear load; or pull-through strength; or power input; or applied load

P_e Euler buckling load

P_m Mean load

P_u Ultimate load

P_y Load at yield strength of alloy

p Pitch of corrugation; or burst pressure

R Resistance of the part; or radius; or stress ratio in fatigue

R_1 Outside radius

R_2 Inside radius

r Radius of gyration; radius between web and flange; or notch radius

r_o Polar radius of gyration

r_x Radius of gyration with respect to x axis

S Section modulus; or calculated strength of the part; or Strouhal number

S_a Section modulus of aluminum part

SC Shear center

SR Stress ratio

S_s Section modulus of steel part

S_x Section modulus about x axis

s Developed length of corrugation; or spacing of transverse stiffeners

s_e Effective width of sheet acting with stiffener

T Torque

TS Tensile strength

T_u Ultimate torque

t Thickness; or plate thickness; or metric ton

t_1 Thickness of flange

t_2 Thickness of web

t_a Thickness of aluminum part

t_s Thickness of stiffener; or thickness of steel part

V Velocity

V_a Load on aluminum part

V_s Load on steel part

W Geometrical parameter for tubes in torsion; or uniform load

W_a Weight of aluminum part

W_s Weight of steel part

w Width of inclined web on trapezoidal section; or width of panel

x_o x distance between centroid and shear center

Y Geometric factor

y Lateral deflection; or amplitude

y_o y distance between centroid and shear center

Z Shape factor for aluminum sections

Z_p Shape factor for rigid plastic material

α_i Fraction of life at stress level σ

β Factor in lateral buckling of beams with unequal flanges

Δ Deflection

δ Crookedness; or out-of-plane displacement

θ Angle of inclination of web

λ Equivalent slenderness ratio; or slenderness ratio

λ_c Equivalent slenderness ratio for torsional-flexural buckling

λ_e Equivalent slenderness ratio

λ_p Equivalent slenderness ratio, plates

λ_s Equivalent slenderness ratio for tube in shear

λ_t Equivalent slenderness ratio, torsion

λ_x Effective slenderness ratio for flexural buckling, about x axis

λ_y Effective slenderness ratio for flexural buckling, about y axis

λ_ϕ Equivalent slenderness ratio for torsional buckling

ν Poisson's ratio

ϕ Factor that accounts for uncertainty in member strength

ρ Unit weight

ρ_a Unit weight of aluminum

ρ_s Unit weight of steel

σ Stress; or axial stress

σ_E Euler stress

σ_a Axial stress; or crippling stress of aluminum part

σ_{ab} Buckling stress under axial loading

σ_b Stress/strength of base material; or failure stress on flange; or bearing strength; or bending stress

σ_{bb} Buckling stress under bending loading

σ_c Critical stress; or buckling strength

σ_f Failure stress

σ_i Stress level

σ_o Buckling strength of simply supported uniform-thickness plate

σ_r Strength of reduced-strength material

σ_s Crippling stress of steel part

σ_t Tensile strength

σ_u Ultimate strength; or ultimate stress

σ_w Failure stress on web

σ_x Stress in x direction

σ_y Yield strength; or compressive yield strength; or stress in y direction

σ_{ya} Yield strength of aluminum

σ_{ys} Yield strength; or yield strength of steel

τ Shear stress

τ_T Shear stress causing yield in web

τ_{cr} Shear buckling stress

τ_w Ultimate shear

τ_y Shear yield strength

Index

ABOUT THE AUTHOR

Maurice L. Sharp is a senior technical consultant for Alcoa Technical Center, Aluminum Company of America. He is chairman of the Aluminum Association Engineering and Design Task Force, which has responsibility for structural specifications for the aluminum industry in the United States.